Mrs Heather Montgomery.
2003.

Breaking new ground

The Bishop's Road, Castlerock
Esler Crawford Photography

Breaking new ground

Fifty years of change in

Northern Ireland agriculture 1952–2002

Derek W. Alexander

Michael Drake

GLENFARM HOLDINGS LIMITED

THE
BLACKSTAFF
PRESS

BELFAST

First published in 2002 by
The Blackstaff Press Limited
Wildflower Way, Apollo Road
Belfast BT12 6TA, Northern Ireland

Design by Corporate Document Services Limited, Belfast
Printed in Northern Ireland by W & G Baird Limited

A CIP catalogue record for this book
is available from the British Library

ISBN 0-85640-713-5

www.blackstaffpress.com

Contents

Preface

Probably no other industry has undergone as much change within the past fifty years as the agricultural industry. Farmers in Northern Ireland have modernised and reshaped their buildings and holdings beyond recognition, and their husbandry methods are as good as those anywhere in the world. They have willingly applied many of the scientific and technological developments that have taken place in the second half of the twentieth century. Methods of husbandry and the structure of the industry have changed, as have the lives of people working on farms.

Two of the most significant advantages of Northern Ireland agriculture over the years have been the ability to grow high-quality grass, and the livestock-husbandry skills of the farming community. The industry's contribution to the Northern Ireland economy and to the food industry over the past fifty years has been immense. From the start of the Second World War, farmers have played an important role in food production. For the first time, farmers in Northern Ireland were accorded the same prices for their commodities as producers elsewhere in the United Kingdom. This had a stimulating effect on the industry. Its inclusion in the United Kingdom price support arrangements was also a major achievement. This proved to be most beneficial for agriculture in Northern Ireland in the years that followed.

Glenfarm Holdings, originally known as Ulster Farm By-Products, commenced its operations at Glenavy in 1952. Although initially involved in the rendering business, the co-operative has expanded into other areas of production. This book celebrates fifty years of the co-operative's existence, and marks the anniversary of its founding by recording some of the history and development of agriculture in Northern Ireland over this period.

Many of the facts and figures on Northern Ireland farming have come from the government's annual farm census, which provides a wealth of information on changes over the years. Statistical information covering the fifty years from 1951 to 2001 has been assembled and used to highlight the industry's trends. The figures and prices over a period of time obviously need to be interpreted with care and not taken out of context. There have also been some changes in definition of census data over the years, and the purchasing power of the pound has decreased considerably, particularly in the 1970s. Non-metric measurements used in the 1950s and 1960s have been retained in cases where they would have been used.

Various publications from the Ministry of Agriculture and its successor departments have been valuable sources of supporting information. The title Ministry of Agriculture for Northern Ireland applied during the 1950s, 1960s and early 1970s. In a reorganisation of the form of government, the title of the Ministry of Agriculture was changed to the Department of Agriculture for Northern Ireland, from 1 January 1974. In December 1999, when power was devolved to the Northern Ireland Assembly, the name was altered further to the Department of Agriculture and Rural Development. The title applying at any particular date has been used as appropriate in this book.

So many events and changes have taken place in the Northern Ireland agricultural industry over the past fifty years that it is clearly not possible to record them all. We have tried to present a balanced picture of the main influences and developments, recording the facts and some of the impacts on the rural countryside. Glenfarm Holdings and the authors hope that this publication will be of interest to members of the co-operative, farmers, those involved generally in the agricultural and food industries and, also, the general public.

The units of measurement have been retained as recorded at the time. Conversion rates are shown below.

Metric Conversion

One acre	0.4047 hectares
One pound	0.4536 kilograms
One ton	1.016 tonnes
One gallon	4.546 litres
One hectare	2.471 acres
One kilogramme	2.205 pounds
One tonne	0.9842 tons
One litre	0.2200 gallons

Prior to decimalisation, the units of currency were pounds, shillings and pence. There were 12 pence in a shilling and 20 shillings in a pound.

Acknowledgements

The authors wish to express their appreciation to the numerous individuals who helped to make this book possible. Initial thanks are due to those in Glenfarm Holdings who conceived the idea for the book and instigated its publication, namely Michael Quinn, Group Chief Executive and Geoffrey Conn, Chairman. Our thanks also go to Douglas Higginson, former Managing Director of Ulster Farm By-Products and to Syd Spence, Group Secretary of Glenfarm, for their contributions to the company's historical background. We are also grateful to Maria Ryan for her co-ordination of this project.

Many people assisted by giving interviews, providing information and commenting on developments in the agricultural industry over the past fifty years. It is probably unfair to single out particular individuals, and others have indicated their preference not to be given any prominence. Nevertheless, for the record, the authors would like to include the following in their thanks: Eddie Conn, Bill Hodges, Hugh Kirkpatrick, John Lynn, Willie McCahon, Robin Morrow, James T. O'Brien, Sam Shaw, Tom Stainer, Harry West, Bob Wilson and Jimmy Young.

Thanks also go to the many organisations whose publications were used as sources during the course of compiling this book and which are acknowledged in the text and bibliography. The authors acknowledge the assistance given by the librarians and staff and the many sources of information which were made available in the Belfast Central Library, the Library of the Department of Agriculture and Rural Development, the Main Library of the Queen's University Belfast, the Agriculture and Food Science Library at Queen's, Downpatrick Library and Ballynahinch Library. The various newspaper archives were also valuable sources of information. We include here particularly those of *Farmweek*, the *Belfast Telegraph* and the *Belfast News Letter*.

Since this is a brief history of the agricultural industry, based on facts over fifty years, the various statistical and other publications of the Department of Agriculture and Rural Development (and its predecessor departments) were absolutely indispensable sources in researching and compiling this volume. The authors would like to record special thanks for the use of the many government census statistics on the Northern Ireland agricultural industry.

The authors also acknowledge all those who supplied and gave permission for the use of their photographs. They are also grateful for the help provided by staff at Blackstaff Press at all stages in the production of the book.

The authors congratulate Glenfarm Holdings on their fiftieth anniversary and wish the co-operative every success in the years ahead.

List of Abbreviations

AHDS	Agriculture and Horticulture Development Scheme
AHGS	Agriculture and Horticulture Grant Scheme
AI	Artificial Insemination
ASSI	Area of Special Scientific Interest
BEMB	British Egg Marketing Board
BOCM	British Oil and Cake Mills
BOD	Biochemical Oxygen Demand, also 'Bull of the Day'
BSE	Bovine Spongiform Encephalopathy
CAP	The Common Agricultural Policy
CJD	Creutzfeldt-Jacob Disease
cwt	Hundred weight (112lb)
DANI	Department of Agriculture for Northern Ireland
DARD	Department of Agriculture and Rural Development
EC	European Community
EEC	European Economic Community
ESA	Environmentally Sensitive Area
EU	European Union
FCGS	The Farm Capital Grant Scheme
FHDS	The Farm and Horticulture Development Scheme
FIS	Farm Improvement Scheme
FMD	Foot and mouth disease
GB	Great Britain
GM	Genetically modified
HLCA	Hill Livestock Compensatory Allowance
IACS	Integrated Administration and Control System
LFA	Less Favoured Area
LMC	The Livestock Marketing Commission
£sd	Pounds, shillings and pence
MAFF	The Ministry of Agriculture, Fisheries and Food
MEP	Member of the European Parliament
MLA	Member of the Legislative Assembly (at Stormont)
MMB	The Northern Ireland Milk Marketing Board
MP	Member of Parliament
NIAPA	Northern Ireland Agricultural Producers' Association
NVQ	National Vocational Qualification
PMB	The Pigs Marketing Board
QUB	The Queen's University of Belfast
RUAS	The Royal Ulster Agricultural Society
SBO	Specified Bovine Offal
SFS	Small Farmer Scheme
UAOS	Ulster Agricultural Organisation Society
UFIL	United Farmers' Investments Ltd
UFU	Ulster Farmers' Union
UK	United Kingdom

Chronology

1946	Inauguration of AI Scheme for cattle
1947	The 1947 Agriculture Act
1952	Registration of Ulster Farm By-Products as a co-operative
1953	End of feeding stuffs control and rationing
1954	Pigs Marketing Board (Northern Ireland) re-established
1955	Milk Marketing Board for Northern Ireland set up
1956	Introduction of Remoteness Grant
1957	BEMB set up
	Introduction of Farm Improvement Scheme
1959	Small Farmer Scheme introduced
1960	*Agriculture in Northern Ireland* first published
	Northern Ireland becomes tuberculin attested for cattle
1961	Seed Potato Marketing Board for Northern Ireland established
1963	Introduction of the Clean Egg Scheme
1965	First importation of Charolais bulls
1967	Establishment of LMC
	Northern Ireland Agricultural Trust set up
1969	Legislation prohibiting sale of horned cattle
1970	Ashton Report on marketing of pigs in Northern Ireland
	Outbreak of Newcastle Disease
1971	Unified Farm Capital Grant Scheme introduced
	Introduction of decimalisation
	Northern Ireland declared a Brucellosis-free area
1973	UK joins the EEC for five-year transition period
1974	FHDS and FCGS introduced
	Farming demonstrations and the crisis in beef prices
	Formation of NIAPA
1978	UK accession to EEC as full member
1980	AHDS and AHGS Schemes commence
1984	Introduction of milk quotas
1985	Introduction of Agricultural Improvement Scheme
1986	BSE first identified in Great Britain
1988	First ESA designated
	AI Services Ltd licensed from DANI
	BSE identified in Northern Ireland
1989	Ruminant feed ban (meat and bone meal)
1990	Ban on Specified Bovine Offal in human food
	French ban on beef imports
1992	Maastricht Treaty signed by EU member states
1993	Single European Market for trade within the EU
1995	Deregulation of milk marketing and formation of United Dairy Farmers
	First Northern Ireland death from vCJD
1996	Health Secretary Stephen Dorrell's statement about likely link between BSE and CJD
2001	Foot and mouth disease outbreak in Northern Ireland

Traditional haymaking at Rathcoole, with all the family involved

Belfast Telegraph

1950s

Introduction

It was the impact of the Second World War and its aftermath that gave Northern Ireland agriculture its biggest boost, firmly establishing it as an important region within United Kingdom (UK) farming. Steps had been taken to ensure that UK agriculture would not again become a hostage to external world circumstances, as had occurred following the 1914–18 war. Joint consultations on agricultural prices were inaugurated in 1945 between the agricultural departments and the farmers' unions. These discussions culminated in the introduction of the Agriculture Act 1947, which assured farmers of reasonable rewards for their efforts.

Hailed by Sir Winston Churchill as 'the bread basket of the nation', Northern Ireland had, during the war years, become the only UK region with surplus food – an invaluable source of supply for other parts of the country. After the war, with priority still on maximum food production, Northern Ireland farmers continued to operate in the spirit of the war effort, maintaining high levels of productivity. 'Wringing the last ounce of food from every acre,' as Churchill graphically said, 'should be the objective of every farmer.' This ethos, combined with other crucial considerations, such as the UK government's desire to reduce high levels of imports, provided the impetus for rapid change and progress throughout the agricultural industry.

At this time, farmers began to take a more active interest in the ownership of processing and other assets. It was as a result of this momentum that some leading members of the Ulster Farmers' Union (UFU) set up the co-operative known as Ulster Farm By-Products, now Glenfarm Holdings. The proposal for a co-operative facility was put forward and approved at the annual meeting of the UFU in May 1952. A rendering factory was subsequently established at Glenavy in 1953.

Much of the factual material in this publication is based on government statistics collected from the farming community. The first agricultural census taken in the British Isles was inaugurated in Ireland in 1847. Apart from one or two short breaks, a census has been taken annually ever since. Until 1953, police enumerators visited farms and recorded the information given to them orally. From 1954, a postal system was adopted in which owners of agricultural land provided details of stock, crops et cetera. The 1947 Agriculture Act gave the Minister for Agriculture power to obtain statistical information from farmers and to impose fairly stringent penalties for non-compliance. Although some farmers may have regarded the provision of this information as a bit of a chore, it was, when taken with agricultural prices and incomes, essential to the development of appropriate agricultural policies for Northern Ireland. Appendices 1 and 2 summarise the census figures for livestock numbers and acreage of crops and grass from 1951 to 2001.

Farm Structure and Employment

The land area of Northern Ireland, including non-agricultural land, is approximately 3.3 million acres. Of these, some 2.3 million acres, or 68 per cent, were under crops and pasture in 1951. This land was farmed by an estimated 58,000 working owners. It is difficult to describe a 'typical' Northern Ireland farmer at the start of the 1950s. However, he was probably an owner-occupier of a small family farm of some 30 to 50 acres, engaged mainly in livestock production on land of only moderate fertility. Owner occupation of farms became possible as a result of a series of UK Acts of Parliament between 1870 and 1925 which enabled tenants to purchase farms from their landlords. The main crops on the 'typical' farm were ware and seed potatoes and ryegrass seed, while eggs, pig products (mainly in the form of bacon), fat and store cattle, milk products, meat and poultry were produced and sold in other parts of the UK. Family farms remained predominant in Northern Ireland during the period from 1951 to 2001.

There had been a decline in the number of agricultural holdings since agricultural statistics were first collected. The drop in holdings of one acre and over in Northern Ireland continued between 1935 and 1960, falling from about 92,000 in the 1930s to 81,000 in 1951, with a further drop to 71,200 by 1961. This represented a decrease of almost 10,000 holdings between 1951 and 1961 – a trend that was to continue as the decades passed. In comparison with Great Britain (GB) where holdings of over 100 acres of crops and grass formed one quarter of the total, the comparable figure was only 2 per cent for Northern Ireland. This illustrates how small the farms in Northern Ireland were in comparison with those in the rest of the UK.

A 'holding' is not the same as a 'farm', since some farmers may own more than one holding, and others may be let in conacre (annual lettings) to other farmers. When a farmer bought or inherited another farm, it was invariably registered with his own farm as a single holding. During the 1950s, the number of farm businesses declined steadily by about 1,000 per year. Excluding non-productive holdings and those entirely let in conacre, the number of farm businesses was estimated at 50,000 in 1961, with an average size of less than forty acres. Consequently, as the decade progressed, there were fewer farmers, and the average acreage of farms was increasing. This was accompanied by a massive decline in the manual labour working on farms, which totalled some 152,000 in 1951, decreasing to about 115,000 in 1961, the largest drop being in full-time farm workers. The increased mechanisation of agriculture, more use of labour-saving equipment and the move away from the mixed enterprises of earlier days to farm specialisation all contributed to this trend. Almost without recognising it, the farm owners of that day were becoming the farm labourers of the future. Appendix 3 provides some details of the farm labour force in 1951 and 2001.

Farmers were encouraged to grow more barley at home to reduce the amount of imported feeding stuffs.

DARD

Agricultural Output

The period from 1951 to 1961 saw a decline of almost 190,000 acres in the area of crops and grass in Northern Ireland. The most notable falls occurred in potatoes, flax and oats. The acreage of potatoes had reached nearly 200,000 during the Second World War, but because of much reduced demand, it had fallen to 144,000 acres in 1951, and continued falling, dropping to 75,000 by 1961. Similarly, the area of flax plummeted from a wartime maximum of 125,000 acres to 21,000 acres in 1951. By 1961, with only 170 acres, the growing of the crop had virtually ceased. Falling prices, together with the unpopularity of the crop with farmers because of its handling difficulties, were the main contributors to its decline.

Concern was expressed in 1952 about the decreasing area of land being ploughed for crops. In 1951, oats was the only cereal of any importance in Northern Ireland, with 316,000 acres in that year, but it was largely displaced by a considerable increase in barley from 3,000 to 111,000 acres between 1951 and 1961. The main reasons for this increase were the availability of more suitable, stiff-strawed varieties, giving better yields, together with easier mechanised harvesting. Most of the barley was fed to the rising number of livestock, especially pigs, because of its lower fibre content in comparison with oats. Barley growing was relatively profitable at this time because of the deficiency payment system, which encouraged farmers to rent substantial acreage in conacre specifically for growing barley. The number of combine harvesters rose from 20 in 1951 to over 900 in 1961, more or less replacing the binder-thresher system of harvesting grain. An increase in the number of grain driers – there were 750 recorded in 1961 – simplified the harvesting of barley, even in wet seasons. The area of potatoes fell by half in ten years to 75,000 acres in 1961. The drop in the acreage of turnips from 12,000 in 1951 to less than 4,000 in 1961 illustrates how this crop declined as a winter feed for livestock. Yields per acre of the principal crops were in the region of half as much again as they had been a century earlier – the ten-year average from 1952 to 1961 for wheat was 22.4 cwt to the acre, barley 23.3 cwt and potatoes 143 cwt per acre.

The total number of cattle on farms in 1949 stood at an all-time high, representing an impressive 30 per cent rise on the figures for 1939 – it climbed again from 961,000 to 1,074,000 between 1951 and 1961. Milk production was considered a priority. The number of cows and in-calf heifers in the cattle-breeding herd was about 331,000 in 1951, expanding to 357,000 in 1961. With the objective of encouraging hill farmers to increase their breeding cattle to provide calves for beef, a Hill Cow Subsidy was introduced in 1953 at £10 per eligible cow. This led to a substantial growth in the number of beef-type cows by the end of the 1950s. There was also a significant rise in sheep and pig numbers in the 1950s. This was a favourable period for sheep, with the population recovering gradually from the drastic consequences of a severe winter in 1947, increasing by more than 75 per cent, from 672,000 in 1951 to 1,183,000 in 1961. During the war, pig numbers had fallen by about 60 per cent because of the shortage of feedstuffs. However, pigs increased again over the ten years to the 1961 figure of 1,033,000, similar to the percentage rise in sheep. This was a reflection of the confidence that had returned to this sector following the restoration of the Pigs Marketing Board (PMB).

More impressive still in the post-war years was the increase in poultry numbers – a huge rise on the 1939 figure of 10.2 million to a peak figure of 24.2 million in 1949. Northern Ireland farms were considered particularly suitable for the production of poultry and eggs. Unfortunately, poultry reached a situation of surplus production with lower prices and a decline in numbers. Total poultry fell from 17.8 million in 1951 to 10.2 million in 1961, with

Table 1: Quantity and Value of Main Commodities: 1944–60

Item	Quantity				Value in £ million		
	Unit of Quantity	1944–45	1954–55	1959–60	1944–45	1954–55	1959–60
Milk	'000 gallons	95,000	107,000	114,000	8.4	14.7	15.7
Fat Cattle	'000 head	176	327	316	3.7	19.8	19.2
Fat Sheep	'000 head	272	535	694	0.8	3.4	4.1
Fat Pigs	'000 head	132	1,394	1,439	1.5	25.6	23.7
Eggs	'000 dozen	53,670	83,580	102,750	8.2	16.9	17.1
Poultry	'000 head	7,575	5,302	6,375	2.5	2.0	2.2
Potatoes	'000 tons	465	506	523	2.4	6.6	5.2
Oats	'000 tons	35.7	9.0	24	0.5	0.2	0.6
Barley	'000 tons	2.8	1	12	0.1	–	0.3

Source: Ministry of Agriculture: Sixth Report upon the Agricultural Statistics of Northern Ireland 1930 to 1953; Seventh Report on the Agricultural Statistics of Northern Ireland 1952 to 1961

reductions occurring in all types of poultry – turkeys, geese, ducks and ordinary fowl. The trends in the quantity and value of agricultural output in the period up to 1960 are illustrated in Table 1.

An important result of the larger-scale and more efficient farming methods of the 1950s was the increased capital requirement of farm businesses. Even a moderate-sized farm required a comparatively large amount of capital. For example, a farm of fifty acres might have had a sale value of up to £15,000 for agricultural purposes, but when the value of stock, machinery and equipment was added, the total fixed and working capital involved might have exceeded £25,000. This was a sizeable figure at the time, and anyone who did not inherit a farm business found it very difficult to become established in a farming career on his own farm. It was also clear, of course, that skill in farm business management was becoming a more important factor in making a success of farming.

Post-war Productivity

As the 1950s dawned, farm productivity levels were exceeding the high standard of the wartime years. Farmers were challenged to meet the government's objective of reducing the high levels of imports by encouraging a larger output of domestic food production. Farmers recognised that production based on imported feedstuffs was inevitably more expensive than using home-grown foods, since the prices of imports were so much higher. The president of the UFU, Arthur Algeo, said in February 1951: 'In view of the feeding stuffs position, the proper course for the farmer in favourable circumstances will be to plough more land, grow as much oats and barley as the area of his land permits, produce fodder beet and increase his silage output.' Grass was described as the 'crop of the century', and the need to get the highest possible output from it became a priority.

Northern Ireland remained an area of surplus food production during the 1950s, while the rest of the UK continued in deficit. Consequently, there was a renewed emphasis on exporting to GB. By 1954, for example, the shipment of livestock to ports in GB was flourishing, with some 246,000 cattle, 310,000 sheep and 327,000 pigs leaving Northern Ireland for destinations in mainland UK. Local produce was fetching good prices in 1951, considering the post-war climate: beef on the hoof was making £6 per live cwt and around £11 a cwt on the hook; hen eggs made almost £7 per case of thirty dozen eggs; farmers' butter was priced at 2s.4d per pound; and a hen could be bought for 7 shillings.

In 1955, farmers in Northern Ireland received an additional boost with the introduction of special financial assistance, which later became known as the Remoteness Grant. This was designed to compensate Northern Ireland farmers who were deemed to be at a disadvantage because of their geographical position and distance from the main markets, resulting in higher transport costs. From 1955 to 1961, the special grant averaged out at £1.3 million per year. The major part of this sum was distributed by way of cattle-headage payments and silage and fodder-crop payments. Obtaining this special remoteness grant gave the Minister for Agriculture, Rev. the Rt Hon. Robert Moore, great satisfaction, as it recognised the peculiar problems facing Northern Ireland in marketing agricultural produce in free-market conditions.

Mr Sam Shaw, a County Down farmer recalls that the Second World War changed everything for farming in Northern Ireland. He remembers that, 'in the 1930s farmers were not getting enough for milk or other products. Milk was being bought from the farmer for a shilling a gallon to be retailed for twice that amount. The obvious solution was to sell milk direct from the farm. Someone advised me to have my cows tuberculin tested. Within a year we were selling all the milk we could produce on our own farm.' However, he says that the 1940s were 'a miserable time, without much social life. Farmers were little more than production-line workers.' In recalling the 1950s, Mr Shaw asserts that they 'brought a respite from the conditions of the earlier years, and were generally much more progressive. We had fifty cows at the time and that was a fair enough number of animals to milk. After the war there was a tremendous demand for food. The sky was the limit and you could produce anything you wanted.' Mr Shaw went on to become a foundation member of the Milk Marketing Board (MMB) that was set up in the 1950s. He is also credited with installing one of the first milking parlours in Northern Ireland.

The 1947 Agriculture Act

The 1947 Agriculture Act marked the beginning of a new era for farming, which had an impact far beyond the 1950s. The Act laid the foundation for agricultural development throughout the UK right up until entry into the European Economic Community (EEC) in 1973. It was of fundamental importance in relation to post-war food production policy. The Reverend Robert Moore stated in a radio broadcast in December 1959 that since the war had ended, 'the most outstanding achievement has been our recognised right to be included as part of the UK in the agricultural price support arrangements under the authority of the Agriculture Act of 1947'. The Act was aimed at 'promoting a stable and efficient agricultural industry capable of producing such part of the nation's food as in the national interest it is desirable to produce in the UK'. This food was to be produced 'at minimum prices consistently with proper remuneration and living conditions for farmers and workers in agriculture and an adequate return on capital invested'.

The Act formalised the annual farm price 'review of the general economic conditions and prospects of the agricultural industry' which took place in the spring of each year. This was known as the annual review, and it became a focus for farmers' interest, arousing detailed comment by their unions every year. It introduced the concept of guaranteed prices and assured markets for individual products. A similar system had been conducted on a non-statutory basis during and after the war, but had been given some permanency only under the 1947 Agriculture Act. A system of guarantees to producers by means of deficiency payments replaced the previous system of fixed prices. In broad terms, therefore, the Act put in place systems that engendered a greater sense of security throughout farming, increased profitability, and so gave farmers confidence to invest in their enterprises.

Guaranteed prices meant that, for the first time, farmers knew in advance what their returns were likely to be, before crops were harvested or even planted. They were assured that the Exchequer would make up any deficit, and so a climate of stability was created. Mr Tom Stainer, chief agricultural economist in the Department of Agriculture and Rural Development (DARD), summed up the benefits of the Act: 'With the coming of the Agriculture Act, 1947, a new "floor" was created for the marketing of agricultural produce. It was probably one of the best forms of agricultural support to come on to the scene.' Further key benefits arising from the Act included the facilitation of improved methods of farm management and husbandry, the provision of equipment and services for pest and weed control and, perhaps most importantly of all, a greater focus on effective marketing. Little wonder then that the Act was quite openly described as the 'farmers' charter'.

Production and Marketing of Agricultural Produce

The resumption of competitive trading after 1953 re-opened the controversial issues surrounding the future of agricultural marketing. Both the UK and Northern Ireland governments favoured producer marketing boards. They were sponsored by the Ministry of Agriculture under the 1933 Agricultural Marketing Act and subsequent legislation. This Act was passed to enable Northern Ireland producers to participate in marketing schemes allowing farmers, acting collectively, to secure equitable prices for their products. One of the most significant developments in the 1950s was undoubtedly the creation or restoration of marketing boards, which fulfilled a useful role in their time. Northern Ireland participated in setting up UK boards for the marketing of wool and eggs in 1950 and 1957 respectively. In 1954, the Northern Ireland Pigs Marketing Board (PMB) which had functioned before the war was re-established. In 1955, the MMB for Northern Ireland was set up; it was perhaps one of the most influential of the boards.

The Milk Industry

Northern Ireland conditions have always been particularly suited to the production of milk. A sufficient supply of rain, averaging about forty-two inches annually over the country, encouraged the plentiful growth of grass, providing a good supply of grazing for dairy cattle. The production of milk for sale off farms in the late 1950s, by some 22,000 licensed producers, was around 100 million gallons per year, reaching 111 million gallons in 1961. The setting up of the MMB in 1955 was of particular significance. Prior to this, the Ministry of Agriculture, acting as agents for the Ministry of Food, had been responsible for the purchase of all milk produced in Northern Ireland. Early in 1954, the UK government decided that the milk marketing arrangements that had been operating since 1940 should cease. In their place, the various Milk Marketing Boards that had been set up were to be restored to the pre-war position. Unlike other parts of the UK, Northern Ireland did not have such a board, and it was decided that the Ministry of Agriculture should continue to control the buying and selling of all milk until alternative arrangements could be made. A year later, on 1 April 1955, the MMB for Northern Ireland was established, with responsibility for the marketing and distribution of milk in the region. This put in place an administrative structure within which the Northern Ireland dairy industry could operate effectively and grow. Productivity levels increased, although some four hundred producers per year ceased milk production throughout the MMB's lifetime.

All licensed Northern Ireland milk producers had to register with, and sell all their milk to, the board. All milk produced for sale became the board's responsibility, including its collection from farms – a major logistical operation. The board provided a full service, covering such items as the purchase of milk cans for sale to producers and the encouragement of up-to-date methods of collection. The role of the Ministry of Agriculture was to ensure reasonable milk-production standards and proper maintenance of premises. The Ministry issued licences to producers whose premises (byres, dairies, et cetera) satisfied certain standards and whose milk passed the statutory Methylene Blue test. The Ministry also licensed distributors of milk. Steps were taken by the board to improve milk quality by introducing a compositional quality scheme, under which differential payments were made according to the total solids content. A bonus was paid for each gallon of superior hygienic quality milk supplied.

The MMB administered the government's guaranteed milk price to producers. Although the board received a higher price for milk sold for liquid consumption than for that sold for manufacturing purposes (for example, for butter or cheese), it operated a pool price per gallon for producers. Milk that went for manufacturing purposes at a lower price obviously reduced the pool price. For the first time, therefore, dairy farmers were sure of receiving the same remuneration for their milk irrespective of the purpose for which it was used. As part of the annual review, the MMB negotiated a financial agreement with the government, specifying guaranteed prices to be paid. Having bought the milk at these prices, it was then the duty of the board to sell the milk to best advantage. The amount that was sold for liquid consumption received the agreed wholesale prices. The amount sold for manufacturing was entirely the board's responsibility, being sold at freely negotiated prices.

One man who remembers the dairy industry of the 1950s better than most is John Lynn, chairman of the MMB for an impressive thirty-three years. He recalled vividly the inception of the board: 'I came in as vice-chairman. A provisional board was set up in 1954 to get us going and then we took over in the following year, under John McAdam as chairman. I followed John into the position of chairman. It was a busy but exciting time. Farmers were all for the board and gave us as much support in those early days as they could.' John Lynn remains firmly convinced that the centralisation of the collection, packing, marketing and sale of milk were fundamental aspects of the board's success: 'Taking over the milk depots which had been built during the war was one of the first things we did. The trade was not too keen on this, but we held out, stating our belief that if we did not get those depots, there would be no board. Statutory powers gave us the right to receive all the milk from the farms with the exception of the few farm bottlers who were allowed to sell their own milk.'

The MMB, while ultimately subject to government control, was careful to enlist the support of farmers, and managed to mediate very successfully between producers and the trade. Among the board's most important achievements in the early days, according to John Lynn, was the increased profitability of dairy farming: 'There was more money in milk for the farmers – and vastly improved productivity levels. When I first came to the board, lactation yields per cow were around the 500-gallon mark, my own farm averaging 550 gallons.' The number of milking machines more than doubled in a ten-year period.

The Pig Industry

A marketing board of sorts for pig production had existed in pre-war days but its activities had been taken over during the war by the Ministry of Agriculture. After the war, the Ministry introduced a farm-to-factory collection system for pigs going for slaughter, replacing the previous system of taking pigs to collection centres for transport to the factory. This resulted in fewer bruises and bite marks, leading to better-quality bacon. After decontrol in July 1954, it was anticipated that surplus pig numbers could prove disastrous for the market. To help deal with this problem the decision was taken to restore the PMB.

The board's main functions were to purchase and market all 'bacon pigs' intended for slaughter. 'Bacon pigs' were defined as 'clean pigs weighing not less than 185lb liveweight or 140lb deadweight'. Pigs under these weights were sold freely, which meant that the board's activities did not interfere with normal transactions in store pigs or in sales from rearing to fattening farms. Although there was a small pork trade, the PMB in effect controlled the marketing of all fat pigs in Northern Ireland. The pigs were sold mainly to bacon curers and processors, with a small number being shipped to GB as carcases.

A sow and her litter at Greenmount Agricultural College

DARD

The healthy state of the pig industry prompted advances in management techniques and technology, which, in turn, ensured further growth. During the 1950s, pigs were moved from smaller holdings into larger units, and there was an overall expansion in pig breeding. Jordans of Moira were leaders in the development of intensive pig production and sweatbox pig housing. In 1953, an importation of Landrace pigs from GB was permitted, and several others followed. J. Pollock of Ballymoney and E.T. Green of Hillsborough hired special aircraft to transport top Landrace animals from Malmö in Sweden. By the late 1950s, improvements in breeding methods, husbandry techniques, disease prevention and slaughtering procedures had brought rapid change. The installation by two bacon factories in the late 1950s of a carbon dioxide anaesthetising plant brought a new, more humane face to the industry. The reputation of Northern Ireland pig breeders was spreading – in 1957, Landrace pigs from prominent breeders such as William McCutcheon in Ballywalter, James Gray of Warrenpoint and George Bryson of Poyntzpass were being shipped to such far-flung destinations as New Zealand.

All of these measures meant that the 1950s was a boom period for the Northern Ireland pig industry, and the number of pigs increased steadily. The bacon factories, whose total capacity had expanded, absorbed the majority of this number and, by 1958, bacon production had reached an all-time high. About 75 per cent of the bacon produced was exported to GB. The board had contributed to the maintenance of a relatively high pig population and an increase in marketing of fat pigs. Over 1.4 million fat pigs were produced in 1959–60 at an estimated value of almost £24 million, despite a steadily reducing guaranteed price. On the whole, the PMB of the 1950s and the vast improvement in the industry for which it was largely responsible were seen as overwhelmingly positive. The majority of producers were greatly appreciative of the board's efforts and the sense in which it consistently championed their interests.

The Egg Industry

Prompted by the other sectors, egg producers were anxious to emulate the success of the MMB. Farmers' unions across the UK came up with a draft proposal for an egg marketing scheme, which was submitted to Westminster and to Stormont. The scheme was approved by both parliaments and, from July 1957, the British Egg Marketing Board (BEMB) assumed responsibility for the sale and marketing of all domestic egg production. Although the board was a UK institution, Northern Ireland was accorded full representation on its main body and its various committees.

Although total poultry had fallen from about 18 to 10 million over ten years, egg production continued to represent an important sector of the agricultural economy. Levels of production rose steadily with the greater proportion of its output being exported. By 1960–61, production was 91 million dozen eggs for consumption, but the price was only 3s.2d per dozen, in comparison with 4s.2d in 1955–56. Egg producers were nevertheless able to look at better husbandry techniques and breeding methods, which, in turn, enabled

Corner of a deep-litter
poultry house

DARD

them to increase productivity even further. During the early 1950s, the use of deep litter was replacing the free-range system of keeping laying hens. Under the built-up or deep-litter system, laying birds were confined continuously to the same house, and the litter allowed to accumulate. The litter was usually peat moss, wood shavings or chopped straw. Success of the system depended on the litter being kept in a dry condition, allowing the bacteria to act on the droppings to create a dry mould. Artificial lighting to encourage egg production during the short winter days was more easily arranged with layers housed inside. Battery cages for laying hens were becoming more frequent by the late 1950s, and the intensive production of broiler chickens for poultry meat was also starting to develop at this time.

One feature of the early 1950s was the continuation of the annual egg-laying tests run by the Ministry of Agriculture at Stormont and later at Gosford. They ran over a period of forty-eight weeks, with six pullets per entry. The most prevalent breeds then were Rhode Island Red, White Leghorn and Light Sussex. In 1952, the White Leghorn breed had the highest hen-housed average, with 181 eggs per hen in forty-eight weeks.

Harry West, a founder member
of Ulster Farm and Minister
of Agriculture in the early
1960s, enjoys a day out at
the Balmoral Show.

Belfast Telegraph

Farming Politics

One of the agricultural 'influencers' during the 1950s was Harry West, who assumed presidency of the Ulster Farmers' Union (UFU) on 12 May 1955. At forty, he was one of the youngest ever to occupy the office. He went on to become Minister of Agriculture. While not occupied on the farm, Harry West was to be found at the forefront of provincial farming politics.

He considered that 'history and indeed the practical and costly experience of most farmers was that their agriculture industry had been "bedevilled" by too much disunity'. He told his thirty thousand members that they should never entertain a return to a pre-war marketing scene when, as weak sellers, they had too often been exploited. As a Unionist MP at Stormont, he was never slow to slam the often 'disappointing' agricultural estimates presented. When Agriculture Minister the Reverend Robert Moore stated that assistance to the industry was to be paid at the rate of £2.2 million in the years 1954 and 1955, Mr West described it as well below expectations, suggesting that something in the region of £7 million a year was needed. It was he who, in the early 1950s, mooted the setting up of a producer-controlled fatstock marketing board for Northern Ireland. Describing the then system of beef marketing as 'uneconomical', he claimed that during glut periods bidding at the auction marts was influenced by buying rings. 'The industry is determined not to be bled white by middlemen as in pre-war days,' he said. His view was that farming prosperity 'would depend more and more on the ability of individual farmers to work loyally together in all ways for the benefit of the industry as a whole.'

Mechanisation and Electrification

Mechanisation, particularly the use of the tractor, had a revolutionary impact on agriculture. Farming had become substantially mechanised during the Second World War, but the advent of the tractor was hugely significant in transforming work on the farm. To spur on production, the government had offered grants to allow farmers to buy tractors and other farm machinery. In 1939, for example, the estimated number of tractors stood at 550. By the end of the war, and largely as a result of a government hire-purchase scheme, that number had jumped to around 7,000. In 1951, 17,740 tractors were recorded in the census, while horse numbers had fallen from around 100,000 in the early years of the century to only 40,000. By 1961, the number of tractors had increased spectacularly to 31,590, and although there were 10,500 horses, they had by then been virtually eliminated from agricultural work.

Practically all farm horses and horse-drawn equipment had been replaced by tractors and related equipment by the early sixties. Enthused by the possibilities of the tractor, early owners often worked in the fields well into the

night, assisted by Tilley lamps. Subsequent models were, of course, soon fitted with proper lights! Much of the drive towards mechanisation was directly related to Harry Ferguson's development of the three-point hydraulic lift system which revolutionised the use of tractors. Ferguson was born in 1884 at Growell near Hillsborough, and developed a mechanical interest at an early age. His contribution to the development of the tractor and the ability to link many farm implements to it made him famous worldwide by the time of his death in 1960. Although the tractor became 'king', much of the former horse-drawn equipment was modified for use behind the tractor. Apart from this, the work of the blacksmith became very limited. All of this contributed to the 'drift from the land' which is still continuing today, helping to improve the overall efficiency of farming.

Horses were no longer a common sight at harvest time – tractors like the pre-war Ferguson pictured below did the work instead.

Belfast Telegraph

Michael Drake

Mechanisation had an impact on all sectors of agricultural production. Between 1952 and 1960, there was an appreciable increase in both rotavators and potato planters, both of which more than doubled. There were large increases in equipment such as seed and fertiliser drills, dung spreaders and fertiliser spreaders. The rising acreage of barley towards the end of the 1950s was accompanied by an increase in the number of combine harvesters. The combine, with two men on board, swept a season's work away in a day. An increase in the number of pick-up balers reflected the change in haymaking methods to baling of the crop direct from the swathe. Some census details on selected machines and equipment are shown in Appendix 4. Mechanisation changed forever the fabric of farming life as machines increasingly took over the work of men and horses. With a tractor-driven plough, a farmer could, with ease, turn over more acres in a day than he could have hoped to accomplish in a few days with a team of horses. There was little sadness at the passing of an era where manual labour had been the norm.

There was considerable progress in the electrification of farms during the 1950s. The Northern Ireland Electricity Board was responsible for supplying the electricity, and farmers, before obtaining a supply, guaranteed to pay a pre-determined sum that included the current used. By 1961, over 24,000 farms had electricity, and some 30,000 farmhouses had a piped water supply.

Haymaking in County Antrim

Belfast Telegraph

Development of Silage Making

The change from hay to silage making was a very significant development. From 1940 onwards, the value of silage was being emphasised by the Ministry of Agriculture, and at least some silage was being made. In May 1951, the Ministry described silage as:

> that rich food made from produce grown on home soil and available to all who wish to have a supply. Well-made silage can be a more nutritious food than the best hay. Silage is a substitute for meal, in the case of cattle stock, and its use made available more meal for pigs and poultry. Thus, when there is an adequate supply of good silage, it is possible, not only to maintain more of these classes of stock, but to lower the cost of maintaining them and of producing milk, beef etc.

The merits of good silage were well-known in the mid-1950s. Silage making was no longer a specialised job, nor a process for which a lot of skilled labour

A typical pit or trench silo

DARD

was necessary. The perfecting of the pit or trench silo brought silage within the means of every stock-owner and made it possible to have good animal food at a low cost. This was most welcome at a time when purchased feeding stuffs were in short supply and their price was up to five times higher than the pre-war figure.

In the post-war period, government grants and subsidies encouraged modernisation and brought a new prosperity to farmers who had been struggling before the war. For example, the grants allowed farmers to build new potato houses as well as buildings to house new machinery. Silage grants also became available to encourage the making of silage as a winter feed. Such was the interest in silage making that, in 1956, the number of silos erected under the Agricultural Development Scheme was about twice the number built during the previous twelve months. The Silage Payments Scheme also provided grants on silage considered of suitable quality for feeding to livestock. In the years 1958 and 1959, payment was made to silage makers at the rate of 20 shillings per ton for the first 100 tons and 12s.6d per ton on the next 400 tons. In 1957, over 400,000 tons of silage was made on Northern Ireland farms, increasing to 830,000 tons in 1961. The method of harvesting silage was also evolving, with the green-crop loader being most prevalent in the early 1950s.

Sam Shaw is reputed to have been one of the people who introduced silage to Northern Ireland. 'I don't know if I was the first to bring silage into Northern Ireland but I was among the first few who pioneered it,' he recalls modestly. Sam saw grass silage for the first time at the Hanna Dairy Research Institute in Scotland in 1939:

> I was intrigued by this, came home and made some on our own farm. We cut the grass, rowed it up and filled a silo with it. Everyone frowned at what we were

doing. It looked like dung and we were looked on as madmen. At that time we fed our cows in winter with turnips and mangolds. We had a big frost and this fodder supply ran out. I went out one night when everyone was away and decided to feed eight cows with the silage. They were very keen on it. I think we had molasses in it. My father was against it. Nevertheless we persevered and the cows were milking like mad! We did so for about ten years before anyone else joined us.

By the end of the 1950s hay, which had been a traditional harvest for centuries, was being replaced by silage, a more sophisticated form of livestock diet. It was essential, however, that its feeding value came up to the required standard. The Ministry of Agriculture was concerned that wet weather during the silage-making season would result in low dry-matter silage, with poorer feeding value. Farmers were advised to wilt young, luscious grass for a day or two before putting it in the silo. Many farmers were making silage for the first time, and the Ministry and commercial firms put considerable effort into explaining the best procedures for silage making and feeding. The traditional system of producing milk during the winter months had been to house the cows in byres and feed them hay and purchased meal. By the mid-1950s, self-feeding with cattle having access to silage was becoming more prevalent. Because silage was a succulent food with a very high moisture content, it was very heavy to work with and cart about. So, it was felt, why not let the animals do the carrying themselves? The cows were kept in loose courts, had continual access to silage, and were milked in specifically designed parlours. All of this contributed to a reduction in the cost of milk production.

Livestock Improvement

Because of the importance of livestock products to the Northern Ireland agricultural economy, the Ministry of Agriculture has had a long-established policy of assisting farmers in improving the quality of their homebred live-stock. Indeed, government intervention in agriculture has been a consistent feature over the years. Various schemes were in operation with the object of stimulating the breeding of high-class stock. A central principle to each scheme was the location of high-grade sires with selected farmers. Under the Cattle Breeding Scheme, annual premium payments ranging from £24 to £40 were paid during the 1950s to farmers who kept premium standard bulls. Participating farmers had to comply with prescribed conditions and make their bulls available for service at a low fee for a period of three seasons. This was of particular value to smaller farmers since the purchase of a good bull was often beyond their means. The annual bull sales held at Balmoral Showgrounds under the auspices of the Royal Ulster Agricultural Society were often important meeting places for those interested in obtaining bulls of the required standard.

Similar schemes were available for Northern Ireland's pig, horse and sheep populations. Under the Swine Breeding Scheme, premiums of £15 were paid in 1959 to selected farmers who obtained approved pedigree Large White boars for local use. A pig litter-testing station was set up at Greenmount Agricultural College to test the progeny of boars. At that time, two tests were

carried out each year, involving four litters from each boar. Although the importance of horse breeding had decreased because of mechanisation, there was still some demand for horses in the 1950s. Breeders were paid premiums on high-quality stallions that were available for service. Similarly, the Ministry aimed to raise the quality of the favoured Blackface sheep population by distributing well-bred rams at reduced prices for mountain farmers – those purchasing such rams under the main scheme were refunded up to half the total cost. The Ram Subsidy Scheme provided financial assistance to breeders wishing to purchase high-class rams or new breeds, particularly from Scotland.

Poultry were not forgotten either, since the Ministry of Agriculture maintained a register of accredited farms. In order to qualify, a poultry farmer had to have premises and equipment of a high standard, and pure-bred stock, and had to follow an approved breeding policy. The purpose of the scheme was to provide a group of farms from which commercial egg producers could buy their laying stock with confidence and which could supply hatching eggs for licensed hatcheries. There were also schemes to assist geese and turkey breeders with eggs being sold from local breeding stations.

Artificial Insemination in Cattle

Artificial insemination (AI) of cattle, again assisted through government intervention, was a key factor in improving livestock genetics throughout Northern Ireland. Owners of smaller herds were able to obtain the service of high-class bulls that would not otherwise have been available to them. Through AI, semen was available from the right type of bull, which had been selected for traits such as conformation or milk yield. AI also proved helpful in preventing the spread of diseases communicable by the bull. The general adoption of AI as the most effective means of livestock breeding contributed greatly to the improvement

A Hereford bull used at
an artificial insemination
centre

DARD

in livestock standards and productivity. The AI service for cattle was introduced in 1946 when 550 first inseminations were made, and the service became increasingly popular thereafter. At the beginning of the 1950s, it was provided from two main centres – Desertcreat in County Tyrone and Riversdale in County Antrim. By January 1952, the whole of Northern Ireland was covered from six AI sub-centres. This meant that the AI service was available to farmers within a radius of fifteen to twenty miles of those centres. The insemination fees charged were £3 for each pedigree and £1.5s for each non-pedigree female.

At the outset, Dairy Shorthorn, British Friesian, Aberdeen Angus, Hereford and Ayrshire inseminations were available. Friesian semen was supplemented by supplies from the Ministry of Agriculture, Fisheries and Food (MAFF). An arrangement was made with the British Friesian Society to enable pedigree females to be inseminated from imported Dutch bulls located at the Scottish MMB's insemination centre in Renfrewshire. By 1959, the Ministry of Agriculture was able to offer herd owners the service of having cows inseminated with semen from bulls of their own choice. The widespread adoption of AI at a national level was relatively rapid – in 1951, 13,000 cattle were artificially inseminated. By 1952, this figure had risen to 19,000, and by 1959, some 96,000 cows were inseminated, representing more than one-third of the cow population in Northern Ireland. Indeed the Ministry's 'AI men' became almost as ubiquitous as the postman as they drove around the countryside on their insemination rounds. As John Lynn has commented: 'Artificial insemination was one of the most significant things to happen to the industry.' Progeny testing enabled farmers to identify genetic traits or merits that could benefit particular breeds or herds, and to identify the best 'milkers'. Through AI, there was a gradual improvement in the standard of cattle in the country.

Controlling Animal Disease

As animal production became more specialised using intensive husbandry methods, the health of herds became vitally important. Maintaining high health standards was an essential pre-requisite for profitability. Strict animal health controls and disease eradication schemes made expansion in livestock numbers possible. The monitoring and prevention of livestock disease was, therefore, an essential activity, and it took on a more organised form in the early 1950s. The government endeavoured to prevent the introduction of disease through rigid control over all livestock importations. Principal diseases targeted by government legislation included bovine tuberculosis, anthrax, swine fever, sheep scab and fowl pest. In most cases, incoming livestock had to be isolated for a three-week period at the Ministry of Agriculture's quarantine station. Even when it was not considered necessary for animals to be quarantined in this fashion – for example, where large numbers of sheep were being imported from Scotland – great care was taken to ensure that permits were issued only for animals whose health background had been thoroughly checked.

The Tuberculosis (Attested Herds) Scheme, which commenced in 1949, encouraged new standards of health awareness. Up to 1949, 40 per cent of dairy cows in Northern Ireland were infected with tuberculosis, resulting in estimated losses of £0.5 million per annum, through loss of milk and meat, and poor thriving animals. At a time when government was calling for increased output and reduced costs, tuberculosis was dissipating profits. It was clearly in the interests of all herd owners that the eradication programme be completed at the earliest possible date. At the outset, the scheme was voluntary and a bonus was payable to encourage farmers to join. By 1952, over 1,000 herds, totalling some 36,000 animals, were certified free of tuberculosis – this represented a three-fold increase in a three-year period, and was hailed as quite a success. The eradication scheme continued to operate voluntarily throughout the 1950s until, by the end of December 1959, approximately 85 per cent of the cattle in Northern Ireland were covered. Compulsory eradication commenced in late 1959 and, by November 1960, Northern Ireland was virtually free from bovine tuberculosis, making the scheme an outstanding success.

The government made stringent efforts to eradicate swine fever, which had been rife throughout Ireland for some time – there had been outbreaks in 1951 and 1956, and again in 1958. On 19 February 1958, the Reverend Robert Moore, then Minister of Agriculture, signed an order placing restrictions on the dealing and movement of pigs. In Ballyarnett, near Derry city, over 100 animals died because of the disease, while, in Newmills in County Tyrone, a grim total of 517 pigs had to be slaughtered. The outbreak affected normal pig auctions as well as the exhibiting of animals at agricultural shows. The restrictions were removed in May 1959. The Swine Fever Act was ultimately, therefore, an effective measure, and stiff fines and court summonses brought home to farmers the seriousness of the unauthorised movement of pigs.

During the mid-1950s, there was an increasing incidence of hypomagnesaemia among cows on improved pasture. This was caused by a deficiency of

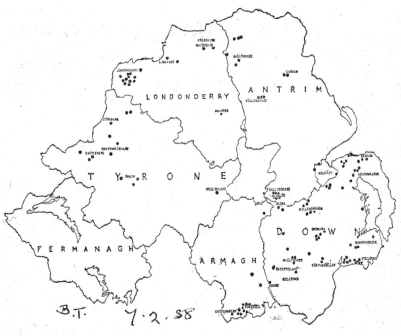

Contemporary newspaper clipping demonstrating the
incidence of swine fever since 1956

Belfast Telegraph

AREAS HIT BY SWINE FEVER

This map shows the incidence of
swine fever throughout Northern
Ireland since the first outbreak
was notified in 1956. In all, 26
cases have been notified, resulting in the slaughter of 15,234
pigs. Compensation amounting
to over £151,000 has been paid to
farmers. The map shows an
almost clear "corridor" running
diagonally across Northern Ireland. Fermanagh is free of the
disease, and South Antrim has an
almost clean bill of health.

magnesium in the blood. The condition came to be known as 'grass staggers'
because affected animals showed muscular spasms and excitability, often when
they were grazing on lush pasture.

Farmers were making increasing use of antibiotics in meal mixtures for
fattening pigs and table poultry. In 1953, farmers were warned about the
dangers of using antibiotics without due care, by feeding them indiscriminately to all classes of animals. From September 1953, specified forms of penicillin
and aureomycin only were permitted in the meal mixtures for fattening pigs
and table poultry. However, in 1958, Dr David Luke, senior veterinary research
officer in the Northern Ireland Ministry of Agriculture, expressed his concerns, when speaking to the Farmers' Club in London, about the level of drugs
being administered to farm animals. These had, he stressed, 'reached astonish-
ing proportions', with 'the unsuspecting pig population receiving more than
its due amount'. Dr Luke hoped that 'poor husbandry methods would never
be sheltered from the economic blizzard by an umbrella of drugs and
antibiotics'.

Some readers may recall the outbreak of myxomatosis in rabbits in the
mid-1950s. The disease occurred amongst wild rabbits in certain districts of

Northern Ireland and spread very rapidly to others. It was a highly contagious and fatal disease. Sick rabbits were often seen wandering aimlessly about fields and ditches, almost blind, with a watery discharge from their eyes. Most rabbits contracting the disease died.

Greenmount Agricultural College, Antrim, where successive generations of students received a thorough grounding in agriculture.

DARD

Agricultural Education

The foundations for agricultural education in Northern Ireland were laid in the early part of the twentieth century. Instructors in agriculture were appointed whose function was to foster better farming practices by visiting farmers, addressing meetings and conducting agricultural classes. Part-time classes for young men already engaged in farming were arranged to explain the principles of good crop and animal husbandry. Residential centres were also established for further education in various aspects of agriculture. The Ulster Dairy School, Cookstown (which was renamed Loughry Agricultural College in 1949), opened in 1908; Greenmount Agricultural College, Antrim, in 1912; and the North West School, Strabane, in 1914. The Northern Ireland Parliament was established in 1921 under the terms of the Government of Ireland Act, 1920, and responsibility for Northern Ireland agriculture passed to the Ministry of Agriculture. The Faculty of Agriculture at the Queen's University, Belfast, was established in 1924.

Pictured above in its original location on Elmwood Avenue, the Faculty of Agriculture at Queen's University Belfast, where much valuable agricultural research was carried out over the years. Newly built facilities were opened at New Forge Lane, Belfast, in 1973.

DARD

During the early 1900s, the pace of scientific research had been gathering momentum and the principles underlying farming practices were becoming better understood. For example, the various minerals which growing plants obtained from the soil were identified, and artificial fertilisers began to replace the natural elements. Livestock was being improved by the application of selective breeding techniques. Government was also encouraging farmers to increase food production by improving their efficiency and yields. Farming was beginning to move towards an organised business, rather than a way of life or subsistence occupation based on the small family farm.

During the 1950s, therefore, there was a growing demand from farmers to improve their knowledge about agriculture. Adopting sound scientific methods could help to achieve higher output, with lower costs and better-quality produce. Farmers wanted to understand the latest scientific developments affecting farming, so that they could make the best decisions, in their own circumstances. It was clear that, if the production objectives of the government were to be achieved, the highest standard of education should be available to those wishing to farm.

Throughout the 1950s, part-time courses conducted by staff of the Ministry of Agriculture formed an important part of vocational training in agriculture. Classes were held at local centres, usually on winter afternoons or evenings, covering general agriculture, animal husbandry and grassland management. Local Ministry advisory staff took the courses so that they could apply them to farming conditions in the area and give emphasis to those enterprises and practices of greatest relevance. One-year certificate courses, covering the general field of agricultural science, were provided on a residential basis at Loughry and Greenmount Agricultural Colleges. These were aimed mainly at students who had the opportunity to become farmers. At that time,

Loughry placed special emphasis on poultry and dairying, while Greenmount emphasised pig husbandry, crop production, cattle management and farm mechanisation. Many certificate holders returned to farming, whilst others took up posts in ancillary industries such as the supply of feeding stuffs. Some also took the course as a preliminary to their university education. At the end of the 1950s, there were plans for expansion of the accommodation at both Greenmount and Loughry. Short, intensive courses of three months' duration were provided at Strabane Agricultural School for young men and women who intended to become farmers.

Students interested in poultry husbandry were able to take the diploma course at Loughry Agricultural College. The courses were designed mainly to provide personnel for services to the poultry industry, and also for those who became production unit managers. Students of the highest academic ability took courses in agricultural science in the Faculty of Agriculture at Queen's University Belfast. They studied either for the general degree of Bachelor of Agriculture or for a Bachelor of Agriculture in certain specialised subjects such as Chemistry, Botany or Zoology. During the 1950s, the majority of students pursued the four-year general course in agriculture. As well as crop and animal husbandry and related sciences, economics and farm management were taught.

The future prosperity of the farming industry depended on having a well-educated farming community. The impact of agricultural education in the 1950s was that the industry obtained vital knowledge and achieved considerable competence to counter the increasing competition from outside Northern Ireland. Again reflecting the new dynamism in agriculture in the post-war years, a great deal of research and experimental activity in agriculture and food processing was undertaken at the various educational institutions. Favoured areas of research included the nutrition of poultry, cattle, sheep and pigs, soil and fertiliser problems and milk testing.

Formation of Ulster Farm

Glenfarm Holdings was originally known as Ulster Farm By-Products. In 1995, the Ulster Farm name was transferred to a new limited company, which had been formed for that purpose, and this business is a wholly owned subsidiary of Glenfarm Holdings – the assets necessary to carry on the rendering business having been transferred to the new company. For the purposes of this history, reference made to Ulster Farm until 1995 means the original co-operative.

Until Ulster Farm commenced its operations, the only rendering company in Northern Ireland was Robert Wilson and Sons, situated on the Moira Road in Lisburn. Its factory was named 'The Burnhouse Factory', after the village in Scotland from which Mr Clement Wilson originated, not because they burned animals – a common misconception! Because Burnhouse had no commercial competition, farmers believed that this situation was monopolistic, leading to unacceptably low prices being paid for fallen animals.

A number of prominent farmers decided to investigate the possibility of setting up an alternative facility. The prime movers were James Megarry

James Megarry Wadsworth, chairman of the Ulster Farm board, 1952–1958

Crawford Wadsworth

An early Ulster Farm share certificate

Glenfarm Holdings

(Garry) Wadsworth, George Ervine, George Fulton, John Martin, Edward (Eddie) Swain, Harry West, Everett Wilson and Verdun Wright. These men became the first directors when the society was formed in 1952, and Garry Wadsworth was appointed chairman. After deciding that a co-operative registered under the Industrial and Provident Societies Act would be the most appropriate organisation, the society was registered in November 1952.

The first board meeting took place in Belfast in February 1953. Mr Percy Smith was seconded from the UFU to act as the society's first secretary. At this meeting the Ulster Bank was chosen as the society's bankers, and remains so at the time of writing. Similarly, James Baird and Company were appointed as the society's auditors and, through their various amalgamations, remain so today (as PricewaterhouseCoopers). Capital for the society was raised at UFU meetings throughout Northern Ireland. By December 1953, some 6,159 members had purchased 76,451 £1 shares. This increased further as the months passed, although the new co-operative had some difficulty in meeting its capital requirements in the early years.

The quest began to identify a suitable site for the rendering business. It eventually came down to a choice between land in the Tandragee area and a farm at Glenavy. The latter was chosen as being the most suitable, mainly because of the sloping nature of the site, which would assist the transfer of raw material down through the various stages of the rendering process. This asset, coupled with the Glenavy farm's accessibility to good roads and a supply of water from the nearby river, clinched matters, and, in March 1954, the board unanimously agreed to purchase Glenconway Farm, Glenavy, from a Mr Nicholson for £9,000.

In addition to the quest for a factory site, research was taking place into the most suitable plant and machinery. During 1953, Garry Wadsworth and Percy Smith travelled to Cahir and Cork and to Scotland to look at knackeries. By September, they were able to report that they had inspected a fat extraction system, made in Denmark. Mr Wyman, who had been appointed engineer to the company in 1954, was sent to Denmark to evaluate the equipment in use. Other plant, including British-made alternatives, had also been considered but rejected. On this evidence, it was decided in February 1955 to purchase Lildal plant for the Glenavy factory. This plant was based on a process known as 'dry rendering', using batch cookers followed by solvent extraction. The cost of the mechanical installation for the whole process line was £43,192. The estimate for the building, effluent plant and site works came to £53,700. When professional fees were added, the total estimated cost was £103,792, on which the society received grant-aid of £25,948.

The plant was pronounced usable during early December 1956. The collection service commenced in the week following Christmas. At this point, the society had twenty-six employees. Initially it was thought that the available fallen animals and butchers' waste bones would be sufficient to return a profit. However, it soon became apparent that this had been over-optimistic. During the first week of operation, the plant processed twenty tons of material. William Gibson, who later became company secretary, informed the board that fifty tons of material per week would be required to meet overheads. By September 1957, the plant was processing one hundred tons per week. This was close to its maximum capacity, so additional processing equipment was purchased.

During the fifties, the society progressively increased its raw material supply, although it was heavily dependent on material from farmers. This necessitated additional vehicles and enlargement of the process facility. In May 1959, Bill Hewitt joined the company as factory manager. He became managing director in 1962, a position he held until his resignation in 1978. This was a most significant appointment, and Bill Hewitt can take much of the credit for turning the business around in its early days, partly with the help of a negotiated loan from the National Farmers' Union Mutual Insurance Society, in which Edward Swain and George Ervine played a major role. Douglas Higginson, who joined the company as laboratory technician in 1959, succeeded Bill Hewitt in 1978 as managing director.

Sadly, the first chairman, Garry Wadsworth, died in January 1958. He was succeeded by Eddie Swain, with George Ervine as vice-chairman – a long-standing partnership of almost thirty years that was to prove most influential in determining the direction of the co-operative. Not only did they make a most important contribution to Ulster Farm, but they were also heavily involved in UFU matters, and were often regular representitives at the annual farm price review negotiations. Appendix 5 records the names of the chairmen, vice-chairmen and board members and Appendix 6 the senior management of Ulster Farm over the fifty years since its registration.

Of Interest to Farmers !

IN the first two years of this Company's operations, we have paid over £22,000 to Farmers for fallen and casualty animals.

This is more than £20,000 greater than would have been paid for the same carcases at the prices ruling before the Company came into being.

Even this, we know, is small compensation for the loss of valuable animals. We do not want them to die, but, if they do, make sure you dispose of them to the best possible advantage by contacting the Local Agent of your own Company.

You will find his telephone number in our advertisements in your local newspaper. Make a note of it for future reference—you can't be sure when you will need it in a hurry. You can also find our depots listed under our own name in the telephone directory. In case of difficulty, your Farmers' Union Group Secretary will be glad to help. Our service covers the whole of the Six Counties.

ULSTER FARM BY-PRODUCTS LTD.

GLENAVY - - - CO. ANTRIM

TELEPHONE : CRUMLIN 401

A co-operative concern owned and managed by Ulster Farmers for the disposal of fallen and casualty animals.

An advertisement for Ulster Farm, placed in the *Lurgan Mail* in April 1959

Glenfarm Holdings

Conclusion

The 1950s marked the beginning of a new era of opportunity for Northern Ireland farmers. A tranche of grants and subsidies brought new prosperity to farmers who had been struggling since before the war. For the first time, they were accorded the same prices as other producers in the UK, by inclusion in the agricultural price-support arrangements, under the 1947 Agriculture Act. This was followed by the 1957 Agriculture Act, which provided long-term assurances and gave much-needed stability to the industry. The result was that increasing supplies of food replaced conditions of scarcity as the 1950s moved on. The control and rationing of feeding stuffs was brought to an end in July 1953. One notable trend was the increase in the barley acreage, most of which was fed to pigs. Intensive broiler production and co-operating within groups was beginning to make progress as the decade ended. The setting up of marketing boards in the 1950s helped to control and direct the marketing of a number of the main agricultural products.

The 1950s saw fundamental changes in the nature of farming in Northern Ireland, with the increasing mechanisation of the farming process, the subsequent massive reduction in manual labour, and the move away from mixed farming towards specialised enterprises. Slurry tanks were starting to replace the middens of farmyard manure that had been a feature on nearly every farm. Farmers were applying more artificial fertiliser and lime to their land to boost crop yields. Silage production was increasing rapidly. Stack yards were disappearing as combines did the threshing of oats and barley in the fields.

New levels of government intervention brought about the standardisation and regulation of the industry, and led to increased efficiency and profitability and to greater co-operation between the various sectors of the farming community. Taken together, the developments began the process by which farms moved from being subsistence-centred to being business ventures. Following the passing of the Minister of Agriculture, the Reverend Robert Moore, on 1 September 1960, the Permanent Secretary of the Ministry of Agriculture, W.H. Long, wrote that during the Minister's period of office (from 1943 to 1960), he had 'made a most outstanding and distinguished contribution towards the welfare of agriculture in Northern Ireland'. Mr Long continued: 'He [the Minister] was deeply concerned about the welfare of Northern Ireland farmers and, because of his great practical knowledge of their requirements, that adequate long-term price and marketing arrangements were vital to the industry.' Agricultural policy during the 1950s was in large measure the direct result of this thinking.

1960s

Introduction

The 1960s were years of intense agricultural growth in Northern Ireland, and the process of modernisation and rationalisation, begun in the 1950s, was maintained. Farmers were moving away from the traditional mixed systems, and farm specialisation meant that many farmers had to develop new skills and inform themselves about the latest techniques. Milking by hand was now a thing of the past, and farmers had to adapt to sophisticated dairy parlours. Livestock farmers had to keep themselves informed about new breeds of cattle that were being introduced, in order to assess their advantages. Silage was playing a greater part in conserving grass for livestock feeding, so farmers were also obliged to familiarise themselves with the latest fertilisers as they tried to grow more and better grass.

There obviously needed to be a concentration of production on larger farm businesses to improve efficiency and lower costs. Increased production, combined with improved productivity and fewer farm businesses, resulted in a substantial rise in the average income per farmer. Eventually, however, food production began to catch up with, and even to exceed, the level of demand. Because of the public expenditure implications, therefore, the government decided that support prices had to be related to the quantities produced. The spectre of taxpayers having to meet an open-ended bill for agricultural production presented political difficulties and, therefore, 'standard quantities' had to be imposed to discourage too much production.

During the 1960s, market outlets had to be identified and supplied against competition from other countries, and quality of products had to match demand. Despite these strictures, in the 1960s, produce from Northern Ireland farms found its way to many parts of the world, as well as to mainland Britain. For example, during 1965–66, shipments of eggs went to Argentina, and carcase meat was exported to Denmark, Egypt, Malaya, Switzerland and the Lebanon. Seed potatoes were shipped to twenty-seven foreign countries, mostly in the Mediterranean area. However, GB remained the major export market and looked likely to remain so in the future.

One of the key policy decisions was the introduction of the Small Farmer Scheme (SFS) in 1959, which determined the course of farm development for much of the 1960s. The scheme was aimed at encouraging land improvements which would result in more efficient food production. The Ministry of

Agriculture offered free advice to farmers with a view to their adopting new and better production methods. Support was also provided through expansion of the agricultural education facilities in Northern Ireland.

Larger Farms and Fewer People

In the past, Northern Ireland agriculture had been characterised by a large number of relatively small holdings, engaged predominantly in mixed farming. This picture altered in the 1950s and was sustained in the 1960s, with farms becoming more concentrated and more specialised. The number of farms with larger acreages increased during the 1960s, as shown in Table 2. Those of less than twenty acres were falling whereas those of over fifty acres were increasing. From 1961 to 1967, the average size of all farms increased from 37 to 46 acres. Of the 46,000 farm businesses in 1963, about 22,000 were smaller than a one-man business.

Table 2: Farm Businesses in Northern Ireland: 1960s

	1964		1967		1970	
	Number	%	Number	%	Number	%
Under 20 acres	10,900	25.2	8,609	21.5	6,745	18.2
20 to 49.75 acres	19,300	44.7	17,933	44.7	16,027	43.2
50 to 99.75 acres	9,900	22.9	9,863	24.6	10,128	27.3
100 acres and over	2,600	6.0	3,047	7.6	3,613	9.8
Rough grazing only	500	1.2	662	1.6	566	1.5
Total	43,200	100.0	40,114	100.0	37,079	100.0

Source: Ministry of Agriculture: Statistical Reviews 1964–1965; 1966–1967; and 1969–1970

Mr Tom Larmour, later Chief Agricultural Officer at the Ministry of Agriculture, wrote in 1964: 'It is important that new generations of farmers should appreciate the difficulties there will be in the future in attempting to make a living from small acreages, or, for that matter, from any acreage, without adequate capital and skill.' He stated further that (in 1964) only 'half of the active farmers in the industry own a business which can provide them with a genuinely full-time occupation and an income above that of a farm worker's minimum wage'. In June 1967, the Minister of Agriculture, Major Chichester-Clark, reiterated this general view, saying that, 'as the economy develops and efficiency of agricultural production rises, it becomes possible and necessary for fewer people to produce the nation's food requirements. This is a world-wide economic fact and we simply cannot insulate our farmers from its effects any more than we could have taken steps to keep all our blacksmiths in business when farmers turned from horses to tractors.' The message for the 1960s was, therefore, fairly clear – farms must become larger and more efficient if their owners were to make a reasonable living.

In spite of increased prosperity for Northern Ireland farmers in the previous decade, the fall in the workforce on farms continued as the 1960s advanced. In the annual June census, the number of working owners was recorded as 47,700 in 1961, falling to 38,700 in 1971 – a decline of almost 900 per year over the decade. During the same period, the number of employed workers, both full-time and part-time, fell from 16,600 to 9,600. When other family members are taken into account, the total number of persons working on Northern Ireland farms, excluding wives of owners, declined by almost 30 per cent from 92,000 in 1961 to 65,400 in 1971 – a huge drop in those depending on the land for their livelihood. The farming industry was relying to a greater extent than ever before on family labour rather than on non-family employees.

Although fewer people worked on the land in the 1960s, conditions improved for those remaining. The Agricultural Wages Board continued to determine minimum rates of wages for agricultural workers. The board also fixed hours of work and maximum deductions that farmers could make from workers' wages in respect of board, lodging, meals and rents. Conscious of the hours being put in by farm workers on the land, the board sought and achieved longer holidays.

The mechanisation of farming moved forward steadily throughout the 1960s, as machinery and equipment continued to take the place of those leaving the land. By 1969, there were over 35,000 tractors on 37,000 farms, while nearly 32,000 farms had a mains electricity supply. There had also been a significant investment in harvesting machinery with, for example, over 1,800 combine harvesters in Northern Ireland by the late 1960s. Investment in tractors, vehicles and equipment in the year ending 31 March 1971 was estimated to be approximately £7 million. Smaller farmers, usually on inviable farms, were attracted to work outside farming in some of the newer industries. This provided better working conditions and invariably better wages. Neither was it uncommon for farmers to give up farming and to lease their land to other farmers to enlarge their enterprises.

It was also true that there was no longer the heightened demand for food and labour that had existed during the war and throughout the early to mid-1950s. By the end of the 1960s, farming provided employment for about 10 per cent of the total number of people in civil employment in Northern Ireland. This figure increased to 14 per cent if some 20,000 workers in ancillary industries such as the manufacture of fertilisers and feeding stuffs, meat processing, bacon curing, egg packing and canning were taken into account. The agricultural industry, therefore, continued to make a significant contribution to the Northern Ireland economy.

The new machinery of the 1960s now forms part of our agricultural heritage, and tractors like this one are demonstrated and admired at vintage rallies across the country.

Samuel Gray

Assistance for Improving Farm Businesses

Many small farmers with potentially viable farms had the opportunity of improving their business efficiency and living standards through government grant schemes. Two of those having considerable beneficial impact during the 1960s were the Farm Improvement Scheme (FIS) and the SFS. It was considered that these measures contributed a firm basis for a prosperous agricultural future for Northern Ireland.

The FIS came into operation in September 1957. Within six months of its introduction, over eleven thousand applications had been received – clearly a scheme welcomed by farmers! The main impact of the FIS occurred, therefore, throughout the 1960s when most of the work took place and its purpose was to grant-aid long-term improvements to agricultural buildings and land. Eligible farms had to 'be capable of yielding a sufficient livelihood to an occupier reasonably skilled in husbandry'. Acceptable items included the construction of farm buildings, making of roads, reclamation of land, installation of electricity, erection of pens for livestock and erection of fences. When approved work under the scheme had been completed by the end of 1973, total grants of over £18 million had been paid on expenditure of over £60 million – a substantial boost to farming in Northern Ireland.

The SFS began in April 1959 and closed for applications in December 1965, when nearly sixteen thousand plans had been approved. Through this scheme, many small farmers with potentially viable farms had the opportunity to improve their business efficiency and living standards. Grants up to a maximum of £1,000 per farmer were available for carrying out an improvement plan to raise the income and earning capacity of the business over a three to five-year period. Admission to the main scheme depended on the acreage of the farm (twenty to one hundred acres of crops and grass) and the size of the business calculated in 'man-days' equivalent (between 250 and 450 man-days). 'Standard man-days' were standards of labour requirements used by the UK Agriculture Departments in the administration of these schemes. When first introduced, a 'standard man-day' represented eight hours' work for an adult male worker, under average conditions.

Farmers were encouraged to adapt their farming techniques and methods of business management to the higher standards demanded by increasing competition. The SFS allowed farmers to put in roadways, improve fencing and modernise stock housing in order to facilitate more efficient production. Ploughing and renovation of grassland, cleaning of sheughs and minor reclamation works were all eligible for grant-aid. The scheme also encouraged the improvement of land by providing subsidies for the application of lime and fertilisers. When the plans had been completed in 1971, over £11 million had been paid in grants since the beginning of the scheme in 1959.

Two further schemes – the Small Farm (Business Management) Scheme and the Farm Business Recording Scheme – were introduced in 1965. They were aimed at encouraging farmers to improve their farms, and were designed to target those who had failed to benefit under the SFS. Specified records were kept over a three-year period and, in the Business Management Scheme, were accompanied by a management programme of improvements based on the records. The schemes depended on farmers recording information that would show how grants had helped to improve productivity. Advisory staff were on hand to help farmers with the process of recording and using information effectively.

By the late 1960s, many farm improvement schemes were well under way. There is little doubt that they had an appreciable impact in strengthening the farming industry to face future competition. In November 1968, the UK government announced selective expansion plans for agriculture, with the aim of raising output and improving farm productivity. At that time, it was felt that the greatest scope for increasing agricultural production economically in Northern Ireland lay in beef cattle, combined with a continued improvement in grassland and in pig production. The government felt that such expansion was necessary to make further savings on imports in preparation for Britain's entry into the European Economic Community (EEC). It would also help reduce the bill that would be presented by adoption of the Common Agricultural Policy (CAP).

Because of technical and economic changes, capital requirements of farms were increasing. 'Fixed capital' was required for investment in land, buildings and equipment; while 'operating capital' was needed for the purchase of livestock, feed, fertilisers, seeds and fuel. Agriculture was becoming an even more capital-intensive industry. It has been said that 'the traditional methods of raising farm capital were to inherit it, marry it or save it – patrimony, matrimony or parsimony'! In the circumstances of the 1960s, these sources were totally inadequate to meet the demands of many farmers.

Banks became increasingly interested in forging a relationship with farmers and in facilitating agricultural investment during the 1960s. They consid-

A bank manager discussing an investment plan and loan facilities with a farmer.

DARD

ered farming as a business, rather than a 'way of life', so there was little sentiment involved. The result was that United Northern Ireland Banks took the initiative by announcing an improved lending scheme for farmers. Another two banks – the Northern and the Belfast – indicated that, in appropriate cases, they would be ready to offer financial help for the purchase or improvement of farms. Repayments would be made over a period of ten, twelve, or even twenty years. This was the first time that a scheme of this kind had been introduced in Northern Ireland. Many farmers who took the opportunity to increase the size of their holdings through the purchase of more land or by acquiring new or second-hand machinery welcomed the move.

Crop Production and Marketing

There was a decrease in tillage (cultivated land) over the course of the 1960s, with a reduction in all crops apart from barley. This was partly the result of the falling away of the post-war demand for food. In 1961, oats was still the dominant cereal crop, with 183,000 acres under the plough, but this represented a radical fall from the 448,000 acres per annum harvested during the war years. One reason for the decrease was that tractors had largely replaced horses, so oats were no longer in such demand. Barley increased rapidly to a peak of 184,000 acres in 1965, replacing oats as the most prevalent cereal. The result was that farmers were taking more land in conacre for barley growing in order to justify the capital expenditure on expensive harvesting equipment. Conacre rents, therefore, became very competitive, and farmers had to think seriously before taking more land to increase their barley acreage. By 1971, the barley acreage was 140,000, with an estimated yield of 27.9 cwt per acre, whilst the estimated oats yield was 20.6 cwt. There was a noticeable investment in harvesting machinery by the end of the 1960s – for example, investment in tractors, vehicles and equipment in the year 1970–71 was £7 million.

The Home Grown Cereals Authority was set up in 1965 to improve the marketing of home-grown cereals throughout the UK. One of its first duties was to introduce a Contract Bonus Scheme whereby growers who registered to deliver barley or wheat to a merchant several months ahead could qualify for a bonus payment on completion of their contracts. The 1967–68 scheme, for example, provided bonuses of eight shillings a ton for contracts made two months in advance, and ten shillings a ton for contracts made three months in advance. The scheme was financed through levies. Better seasonal spread of barley marketing was also encouraged. Farmers selling barley from July to October suffered deductions in their payments, while premiums were paid to those selling in the December to June period.

The government was still anxious to encourage an increase in barley growing. In 1968, the Ministry of Agriculture wrote:

> Barley growing is attractive for two reasons. First, it can be a profitable cash crop. Secondly, each ton produced in Northern Ireland helps to reduce the burden of import costs to the industry. Within the limits of economics on the individual farm, it is in our interests to grow as much barley as possible.

The Ministry of Agriculture, acting as agent for MAFF, administered the Cereals Deficiency Payments Scheme which, in a five-year period, averaged over £2 million per year. Guaranteed prices, usually referred to as standard prices, were determined by government as part of the annual review, based on standard quantities of national production. When the standard quantity was exceeded, the rate of deficiency payment was reduced proportionally. Also, if the average 'at farm' price of oats and barley fell short of the standard prices, the deficiency was converted to a rate per acre, and an acreage payment made to farmers. Meanwhile, the spring of 1962 was recorded as one of the worst for leatherjacket (cutworm) damage to cereal braids in Northern Ireland for many years – the braids being the first young shoots to emerge from the ground. Poison

bait of Paris green mixed with bran was one of the treatments, although DDT and BHC were also being used.

The potato acreage in Northern Ireland declined generally over the course of the 1960s to less than 25 per cent of the level during the war – only 42,000 acres were grown in 1971. This resulted from increased production in GB and decreasing demand for potatoes. Some 18,000 acres were grown and certified for seed purposes in 1971, with the remainder going for ware – that is, they were sold to the general public for consumption, rather than as seed for planting. Taking the potato crop as a whole, the most prevalent variety was Arran Victory with 9,900 acres, followed by Dunbar Standard with 6,800 acres, Kerr's Pink with 5,400 acres and Arran Banner with 4,000 acres. The government guaranteed that the growers' average price of ware potatoes sold for human consumption would not fall below the standard price fixed at the annual review. To achieve the target price, the Ministry of Agriculture, when necessary, purchased at least part of the surplus for resale as stock feed.

A scheme constituting the Seed Potato Marketing Board for Northern Ireland came into operation in July 1961. The board aimed to bring some degree of stability to the industry, both in producer prices and in maintaining and extending the market. It became the sole buyer of certified seed potatoes. The actual marketing arrangements between grower and shipping merchant were the responsibility of the board, whose functions included publicity and the development of overseas markets. In 1961, the main seed varieties exported to Europe were Arran Banner, Up To Date, King Edward VII and Arran Consul. The main varieties shipped to GB were King Edward VII, Majestic, Arran Pilot, Ulster Prince, Ulster Premier and Ulster Chieftain. The board remained in operation throughout the 1960s, guaranteeing growers of the main seed varieties an assured market for their crops. In 1969, Mr John Clark

Seed potatoes being picked from a sorter

DARD

Loading milk at the depot
of Bangor Dairies

DARD

from Dervock was awarded the OBE for over forty years' work on potato
breeding, leading to the many 'Ulster' varieties – a great service to agriculture
in Northern Ireland.

Milk Production and Utilisation

Dairy farming went from strength to strength during the 1960s, largely
because of the excellent grass-growing potential in Northern Ireland, together
with improved grassland management. At the beginning of the 1960s, around
500,000 acres of grass were being harvested for use as feed – a huge increase
from the pre-war position. With the average size of farms in Northern Ireland
approaching forty acres in the early 1960s, many were becoming large enough
to sustain what was then regarded as a reasonable-sized dairy enterprise. The
production of milk for sale from farms in the year ending 31 March 1961 was
in excess of 111 million gallons. However, further expansion was to follow
throughout the decade, with production figures reaching almost 160 million
gallons by 1971. The 1971 census recorded 215,000 dairy cows, almost the
same number as in 1961.

The estimated average milk yield from the dairy herd improved from about
685 gallons per cow in 1963–64 to 810 gallons by 1970–71, although the
increase was beginning to slow down a little. Over the same period, the
average value of milk increased gradually from 2s.7.5d (equivalent to about
13p) to 16.08p per gallon. This was in comparison with the position applying
in the mid-1950s when the average price of milk per gallon was 2s.11d (equivalent to 14.58p).

With the help of the Milk Publicity Council's advertising campaign, sales of milk for liquid consumption rose in the early 1960s after a period of stagnation in the late 1950s. Liquid consumption amounted to 38.4 million gallons in 1961, rising to 44.5 million gallons by 1971, but this did not keep pace with the huge increase in production. Over two-thirds of the milk produced still had to go for manufacturing, which attracted a lower price. Butter was the least profitable manufacturing outlet. The higher the proportion that went for manufacturing, the lower the price to the farmer. The manufacturing of milk products was, therefore, an important consideration for the milk sector. The amount of milk used for the various milk products in 1958–59 and 1969–70 is shown in Table 3. This demonstrates that as the 1960s progressed, milk production increased, but the proportion going to the most lucrative liquid market decreased. This was not a good sign as far as the price of milk to the farmer was concerned.

Table 3: Utilisation of Milk Produced in Northern Ireland (Million Gallons)

| | Y/E March 1959 | | Y/E March 1970 | |
	Amount	%	Amount	%
Liquid Consumption	36.5	36.1	43.8	28.1
Separated for Butter	22.6	22.3	46.4	29.7
Dried Whole Milk	15.6	15.4	3.1	2.0
Condensed Milk	14.7	14.5	14.8	9.5
Cheese	5.6	5.5	20.2	13.0
Cream	4.0	4.0	18.4	11.8
Other Milk Products	2.2	2.2	9.2	5.9
Total Sales off Farms	101.2	100.0	155.9	100.0

Source: Ulster Year Book 1960–1962; and 1971

The quality of milk was a growing concern for consumers. In response, those controlling the distribution and marketing of milk introduced quality control procedures. The Methylene Blue Test, which gave a measure of the standard of a producer's methods, had been used by the Ministry of Agriculture to assess the keeping quality of milk. Those producers who consistently supplied milk of excellent quality were awarded Superior Hygienic Certificates (introduced in 1963) and were paid a bonus. Strict control of the licensing of milk producers and distributors was maintained and, by March 1970, there were slightly more than 15,000 milk licence-holders. Of these, almost 11,000 held the Ministry's Quality Certificate, which earned them a price premium on sales. There were two types of milk producer's licence – a grade 'A' licence (tuberculin tested), which later became a 'Farm Bottling' licence; and an ordinary milk licence. Producers licensed as grade 'A' put their milk into retail containers on the farm and were the only people allowed to sell raw milk. All other milk was pasteurised before it was sold to the public.

In the early 1960s, the MMB introduced a scheme for the purchase of milk and payment to producers on the basis of compositional quality. Rates of payment varied according to the solids–non-fat content of the milk. Deductions were made for low butterfat content – high butterfat content was more valuable for manufacturing purposes. When a producer's methods were found to be unsatisfactory, the Ministry attempted to improve them through advice, education and steady pressure. Those few licence-holders who were not able to maintain the necessary minimum standards had their licences withdrawn.

One of the problems was that milk was not produced evenly over the course of the year. The amount tended to be higher in summer and lower in winter, making it difficult to achieve a steady workload for those involved in the manufacturing of milk products. As a result, the MMB offered higher prices for winter milk than for summer milk. This policy was successful during the 1960s in flattening out the peaks and troughs of milk production.

Slatted floors in a cubicle house, contributing to savings in labour and bedding costs

DARD

Grassland management had also been changing over the years. In earlier times, a system of straightforward rotational field grazing had been used. With paddock grazing, the fields were divided into a number of paddocks of about the same size. Electric fencing was often used for this purpose and the paddocks were grazed in rotation. There was then a rest period of about three weeks before the paddocks were grazed again. Paddock grazing for dairy cows became widespread throughout Northern Ireland in the late 1960s. With two-sward management, one area was kept for cutting and another for grazing. These improved methods of management helped to raise the productivity of grassland during the 1960s.

Progress in Livestock Production and Marketing

While the government was encouraging beef production, farmers were beginning to question the profitability of their traditional methods. Beef cattle fattened at about two years of age did not yield the best profit. Alternative methods of intensive beef production were being considered. Slatted floors for fattening beef cattle first appeared in the 1960s. The arguments for slatted floors included the reduction in the amount of straw or other bedding material required, saving in labour, better general health of the animals and more economical live-weight gain. 'Barley beef' also appeared in the 1960s – this entailed housing animals inside for all of their lives and feeding them exclusively on concentrate feeds, mainly barley and protein concentrate. The economics of this were, however, greatly influenced by the cost of the calf and barley at any particular time.

The 1960s saw the introduction to Northern Ireland of Continental beef breeds of cattle, the first being the Charolais. This was a major development in beef production, one of the aims being to produce better-quality calves from dairy cows where herd replacements were not required. Semen from Charolais bulls was available from the AI service from 1965. It was claimed that the Charolais had a number of advantages over traditional beef breeds such as Aberdeen Angus and Hereford. These included better growth rates, less fat, higher killing-out percentages and better development of the hindquarters. From 1965, the number of Charolais inseminations rose rapidly to over 24,000 in 1968–69. At the end of the 1960s, farmers were making careful comparisons of this breed with other beef bulls for crossing purposes. Further Continental breeds, such as the Simmental and Limousin, were to join the Charolais at a later date.

During the 1960s, there was renewed thinking, at farm level, in relation to the export of store cattle for finishing in England, mainly in Yorkshire. This prompted the development of a number of privately owned meat plants and abattoirs, and a move towards the exporting of carcases rather than live animals. Exporting carcases proved to be a much more lucrative enterprise than exporting cattle on the hoof, and helped to expand Northern Ireland's meat export trade. Figures show that beef carcases leaving Northern Ireland rose from 15,374 in 1962 to almost 23,000 the following year. By 1964, livestock export figures were proof that the overall livestock industry was booming – beef carcases now stood at 58,225, while pork numbered 16,925, and mutton 176,227. The 1964 Meat Shipping Regulations covered ante- and post-mortem veterinary inspections, hygiene in abattoirs and processing premises and the transportation of meat. They produced high standards of hygiene and handling conditions that enabled Northern Ireland to take up a competitive place within the world meat market.

June 1966 saw the publication of a White Paper on fatstock marketing in Northern Ireland; the White Paper was extensively debated at the time. This subsequently led to the formation of the Livestock Marketing Commission

The introduction of the Charolais breed from France heralded the importation of further Continental breeds, making a significant contribution to the livestock industry in Northern Ireland.

Derek W. Alexander

(LMC) in August 1967. Its objectives were to take steps to reduce the differential in prices for fat cattle between Northern Ireland and GB and to examine the industry in general. In effect, it was not a marketing board but an advisory body charged with functions such as advising the Ministry of Agriculture on the development of fatstock marketing and carcase classification. In addition, it was responsible for promoting the idea of high-quality beef products from Northern Ireland and arranging contracts between meat plants and producers. It also provided improved market intelligence, including the advance weekly publication of prices payable by meat plants. The LMC was chaired initially by Mr Arthur Lovesey; he was succeeded for a number of years by Mr Jack Swinson who was later knighted for his services. The commission drew the attention of producers to the importance of producing good-quality hides so that the best possible price could be obtained for them. The LMC proved to be an extremely valuable body; it still remains in existence under the new title of the Livestock and Meat Commission.

To encourage the production of beef carcases of good conformation and fleshing, without being over-fat, a Carcase Quality Bonus Scheme was introduced in the autumn of 1967. The scheme was designed to help farmers produce the type of carcase required by the market. A bonus was payable on the better grades of beef carcases certified under the Fatstock Guarantee Scheme. Grading was based on carcase characters such as general conformation and fleshing, fat covering and weight. The top grade earned a bonus of £2.5s on each carcase, while £1.10s was paid on the second grade. The lowest grade did not earn a bonus. This acted as an incentive to produce animals that were not too heavy or over-fat.

James T. O'Brien, first chief executive officer of the LMC, believed that producers had to be in a position of strength in the market place and be dedicated to adding value to their raw product. Traditionally, Northern Ireland had been an exporter of live animals. However, James O'Brien believed that 'if value was to be added to the animals, they had to be slaughtered in Northern Ireland and sold in carcase or joint form'. Moreover, he added that if there was to be 'a more equitable balance between sellers and buyers of the raw material, there had to be more competition for it'. Accordingly the strategy adopted was to encourage the introduction of a number of competing packing plants – a strategy that was bolstered by the 'seasonality' of production, which worked against the economic provision of live-cattle shipment facilities. To reinforce competition, a vigorous and sustained publicity programme was launched to increase demand, especially in new areas such as on the Continent. The carcase trade flourished and O'Brien attributes this success to the professionalism and energy of Brendan McGahan and other colleagues in the LMC. 'The success of this policy cannot be challenged,' he asserts.

By the time I retired in 1983 some 96 per cent of cattle disposals from Northern Ireland were slaughtered here and only some 4 per cent shipped live, almost all store cattle. Beef and sheep numbers were at record levels and beef from Northern Ireland was known and appreciated throughout the Continent.

The animal-welfare lobby was becoming more vociferous about the treatment of cattle. Welfare and economic considerations made it desirable that cattle should be dehorned, preferably in the first few weeks of life. Legislation was enacted in 1965 to disallow the sale of horned cattle at a future date, subject to certain exemptions. After more than three years of publicity, it became illegal to sell horned cattle from February 1969. Most farmers selling cattle complied with the law although there were a few who failed.

New arrivals at the Ministry of Agriculture's Northern Ireland Pig Testing Station in Antrim

DARD

Sow stalls and slatted floors

DARD

Developments in the Pig Industry

A high point for the pig industry in Northern Ireland was reached in 1964–65 with a peak in the pig-breeding herd, sows and gilts. However, pig prices had been falling, reaching an all-time low in June 1965. The rising cost of feeding stuffs also made pig production less attractive. These factors marked the beginning of a long-term decline in the pig industry. All bacon-weight pigs continued to be sold by producers to the PMB which, in turn, sold them on to bacon curers. The board was in receipt of Ministry of Agriculture payments under the Fatstock Guarantee Scheme, and the prices paid to producers included these payments. The PMB's prices were also related to a grading system designed to encourage production of the type of pigs best suited to the current market demands. A Pig Production Development Committee was set up in 1964 to monitor and improve the breeding quality of pigs.

There were many changes in the design of piggeries over the years, the trend in the 1960s being towards keeping even more pigs on a smaller floor area. In some cases, this was achieved by having slatted dunging passages in the house, thus eliminating the need for outside yards. With the greater stocking density, insulation and ventilation of the houses became more important. The 1960s also saw a significant change in the way in which pigs were killed. From 1963, all animals slaughtered for human consumption had to go through licensed slaughterhouses. The capacity of the bacon-curing industry was adapted to meet the increased throughput of pigs and to market a substantially increased supply of bacon. The main market continued to be the rest of the UK, which constituted over 85 per cent of total sales of bacon and ham produced in Northern Ireland. The Ulster Bacon Agency played a useful role in helping to market bacon in GB, where there was keen competition from Denmark and other countries.

Eggs and Broilers

Egg production maintained its position in the Northern Ireland farming industry, being worth £14 million a year to producers at the beginning of the 1960s, rising to over £21 million by 1971. The average size of the laying flock had reached 8.8 million birds in 1970–71. Northern Ireland still exported a large proportion of its eggs, contributing about one tenth of the requirement for

Egg producers were encouraged either to produce clean eggs or to dry clean them as a means of maintaining quality.

DARD

eggs in GB. Egg marketing remained in the hands of the BEMB. Through merchants who acted as its agents, the board purchased all first-quality eggs and implemented the government's price guarantee system. Producers owning more than fifty head of poultry had to register with, and sell their eggs to, the board if they were to receive the guaranteed price. Some of the first attempts at contract egg production took place in 1966 when the board operated a Voluntary Contracts Scheme under which producers received up to 4d per dozen more for eggs supplied under contract. Moy Park and Associated Egg Packers were major players in the egg production business of the 1960s.

The egg industry in Northern Ireland was given a huge boost in April 1963 when the BEMB introduced a Clean Egg Scheme. Soiled eggs were one of the major producer problems as consumers naturally preferred clean eggs. Tests had proven that the cleaning of eggs by wet methods caused rapid deterioration,

Newer hybrid strains of poultry and improved management in battery laying cage units resulted in improved yields per bird.

DARD

especially in warm weather. The Ministry of Agriculture exhorted farmers in the following terms: 'Eggs are clean when laid and every producer should try to keep them so.' Eggs that were washed or wet treated received a lower price than those naturally clean or dry cleaned. It proved a real incentive to producers, encouraging them quickly to adopt dry cleaning methods. The scheme resulted in practically all eggs marketed in Northern Ireland being unwashed and, therefore, having good keeping quality. This was essential as 80 per cent of them were sold in GB in competition with eggs that had not as far to travel. The demand for eggs from Northern Ireland improved as the 1960s moved on, injecting new vigour into the industry.

Many new battery laying cage units were installed from 1964 onwards, mostly organised within integrated groups covering both production and marketing. One novel aspect of the cages was that they incorporated nipple drinkers for the supply of water to the hens. Over the course of the decade, it is estimated that average annual egg yield per bird increased from 179 in 1961 to 218 by 1970–71. This was largely attributable to the introduction of newer hybrid strains of poultry and better flock management. In 1966, nearly 1,000 farms were equipped to take delivery of feeding stuffs direct from the supplier's lorry into bulk feed-storage bins – a number that increased rapidly to 2,800 in 1972. Many of these were poultry and pig farmers.

Adjusting the drinker in a broiler house fully equipped with automatic feeder and environmental control

DARD

Although commercial broiler chicken production began in Northern Ireland in 1958, it was during the 1960s that it began to make strides. 'Broilers' were specially bred meat birds of ten to twelve weeks of age, weighing three to four pounds live weight. In the early days, old buildings were converted, but newly designed houses, usually for six thousand birds, began to appear around the countryside. Considerable capital was necessary in setting up new production units. Therefore, some co-operation with groups in the provision of housing, equipment, and feed was advantageous to overcome the shortage of capital. This co-operation, which often also extended to the hatchery and to the processor, was building up a head of steam in the early 1960s. Broiler producers joined with others in integrated groups, using reputable stock and specialised housing. The birds were bought and sold under contract, providing a steady flow to the processing plants. The principal market for most of the poultry meat was in GB. Unlike eggs, where there was a guaranteed support price, poultry meat had to find its own price level in free competition with other foods. The Moy Park and O'Kane groups were the major Northern Ireland firms involved at this time in the production and processing of broilers.

An advisor from the
Department of Agriculture
offering practical assistance

DARD

Agricultural Education and Training

Such were the changes in agriculture that in the 1960s the Ministry of Agriculture strengthened its free advisory service to help farmers keep pace with developments. Those involved in the service were qualified in agriculture, poultry or horticulture, and worked from special advisory centres in each of the six counties of Northern Ireland. The advisory service offered guidance on business management and farm accounting, and on making the best use of available resources, as well as offering help with technical and husbandry problems on the farm. It was also the role of the service to help keep farmers abreast of the results of research and new developments.

Farmers were encouraged to adopt a business approach by keeping farm accounts and updating them regularly.

DARD

There were four main channels of communication between the advisory service and farmers – public meetings, farm visits, advisory literature and the press, radio and television. Halls, schools and other public buildings were used for meetings and discussion groups throughout Northern Ireland. These discussions all proved helpful in getting new ideas established. It was accepted that farm visits were very useful and that certain problems could be resolved only by an on-the-spot call with the farmer. However, such trips were expensive and there was a drive to make better use of advisory correspondence, press, radio and television as ways of communicating and conveying information to farmers.

Those thinking of going into farming were encouraged to take advantage of training, and advice was offered on the various courses. In winter, when farmers may have had more time because of the dark evenings, the advisory service organised part-time agricultural classes in various venues. This drive towards educational training during the 1960s was supported by an expansion of Greenmount, Enniskillen and Loughry Agricultural Colleges. Student capacity at Loughry and Greenmount was expanded to a total of 240 places, and a new college at Enniskillen, with a further thirty places, had its first intake of students in October 1967. In the mid-1960s, emphasis was placed on the keeping of farm accounts, as a means of showing whether the farm business made or lost money. Unfortunately, there was no short cut to dealing with extra paperwork, and farmers were often dilatory about keeping records. The Ministry of Agriculture organised many farm accounts classes and provided

Students receiving practical instruction in handling farmyard manure

DARD

farm management account books especially for use on the farm.

Each year, the Ministry published some one hundred leaflets covering practically every aspect of farming. The journal previously called the *Monthly Agricultural Report* was renamed *Agriculture in Northern Ireland* in September 1960, and was made available free of charge to farmers. It was the Ministry's principal medium for the spreading of information on all technical and policy matters. It contained articles and up-to-date information on agricultural developments in the country. Warnings were given on everything from potato blight to apple scab! For the first time, advertisements were included in the publication from 1960, although the Ministry did not accept any responsibility for their content. BBC farming programmes, *Farmweek* and agricultural sections in the *Belfast Telegraph* and the *Belfast News Letter*, together with the provincial press and literature from commercial firms, all added to the volume of information available to farmers. In addition, *The Farmers' Journal*, the official organ of the UFU, had been published since 1920 as a means of informing UFU membership, legislators and decision-makers about UFU views. From October 1972, the journal was published as part of *Farmweek*. *Farming Life*, which was in its thirty-ninth year in 2002, continues as a supplement to the *Belfast News Letter*.

Advisory, teaching and research staff also played a major role in radio and television broadcasts. As Jonathan Bardon points out in his history of the BBC in Northern Ireland, 'The Ministry of Agriculture … was the first government department to realise the potential of broadcasting and seized the microphone to promote improvement.' Programmes such as *For Ulster Farmers*, *Ulster Farm* and latterly the hugely popular *Farm Gate* became important vehicles for disseminating information and keeping farmers abreast of new techniques and methods. In the 1960s, one of the popular broadcasts was *Ulster Farm Top County*, an inter-county knock-out quiz competition for teams of young farmers, invariably chaired by Mr George Shannon, then principal of Loughry College.

Influence of Commercial Companies

Agricultural development was also assisted by the progressive attitude of leading commercial companies who provided goods and services for farmers. Richardsons Fertilizers was one such company. It had already made a major contribution to farming in Northern Ireland in 1954 when a granulation plant was added to its factory. This introduced granular compound fertilisers to farmers who until then had been using powdered matierals. In 1962, the firm opened a new factory in the harbour area of Belfast at a cost of £2 million, built to take advantage of the most recent advances in fertiliser manufacture. The process was based on mixing mono-ammonium phosphate with varying amounts of sulphate of ammonia and muriate of potash, to produce concentrated fertiliser. The two-sward system, involving separate grazing and silage areas, was encouraged by Richardsons. It had a big influence in simplifying the approach to intensive grassland utilisation and, in turn, had a major impact on dairy and beef farming.

Another company that played a crucial role in developments in the 1960s

Richardsons Fertilizers
factory, including the new
extension opened in 1962

Belfast Telegraph

was the British Oil and Cake Mills (BOCM). The company did not actually begin operations in Northern Ireland until 1951 and, until the opening of its Associated Feed Manufacturers' mill in York Road, Belfast, in 1960, all of its products had to be transported from GB. A second landmark in the company's career occurred in 1961 with the opening of the Templepatrick Fair in County Antrim. Here, over a number of years, the fairs catered for those interested in both pigs and poultry. The company ran public pig performance tests that afterwards played an important part in the BOCM pig-breeding scheme. Both BOCM and Richardsons Fertilizers operated comprehensive advisory services and built very strong relationships with leading farmers who assisted in promoting the effectiveness and success of their methods and products.

Co-operatives

Even in the 1950s, Northern Ireland farmers, faced with large-scale buying organisations, recognised that they occupied a relatively weak position in the marketplace. One option was the formation of more agricultural co-operatives, as farmers realised that pooling of their resources created a much stronger, more effective and more competitive group. The 1960s saw the formation of a number of farming co-operatives and the consolidation of a number of others, such as Ulster Farm, which had been formed in the 1950s. The amalgamation of farms into more viable units and the formation of farmers' co-operative groups were favoured. This trend towards fewer and larger units was looked on by the Ministry of Agriculture as desirable. The Central Council for Agricultural and Horticultural Co-operation also encouraged farmers to join groups in an effort to reduce capital and production costs. The council was prepared to provide grant-aid to its members, provided that they gave a commitment to the group, had a common grassland and fertiliser policy, and shared both machinery and labour. For marketing co-operatives, there was a grant for building and fixed equipment, and for working capital in the case of new co-operatives.

Bob Wilson spent almost forty years with Fane Valley Co-operative Society and, from 1951, saw the co-operative movement operating at close quarters. Fane Valley was then engaged mainly in bottling milk in one-third pints for schools, and Bob describes it as 'a troubled ship since there was no finance for capital investment'. In 1950, there were 253 shareholders, average net profits over a ten-year period were around £2,000, and the intake of milk approximately 65,000 gallons. A short time later, however, net profits were almost hitting the £6,000 figure, and milk intake had gone up by over 110 per cent. He recalls that 1963 was a crucial year for the co-operative:

> Armaghdown Creameries, one of the largest milk centres in Northern Ireland, came onto the market. Fane Valley, together with Leckpatrick and Killyman co-operative societies, acquired ownership. It was a major financial undertaking for Fane Valley. Over the years the co-operative expanded and diversified its business. It provides a good example of a successful co-operative, which was able to sell its goods competitively in a world market.

In spite of these successes, however, the co-operative movement in Northern Ireland was not well supported. The alternative means of strengthening the position of farmers, and the road which was largely followed in Northern Ireland, was the establishment of statutory marketing boards.

Animal Disease and Husbandry

The requirement to notify certain diseases to the Ministry of Agriculture, and Northern Ireland's comparatively isolated geographical position combined to keep the farming industry relatively disease free during the 1960s. The early part of 1960 was marred, however, by the occurrence of several cases of anthrax, some of which were not diagnosed until the animals reached the knackery. Anthrax was a notifiable disease and farmers had a responsibility to report suspected cases. Infection may have been associated with imports of feeding stuffs from abroad that had been infected with anthrax spores. Foot and mouth disease (FMD) and swine fever stayed out of Northern Ireland over the 1960s. However, there was great concern in 1967 that the serious outbreak of FMD in GB during the autumn of that year would spread to Northern Ireland. Public gatherings were cancelled and, through restrictions on importation of possible carriers and movement of livestock, the disease did not reach Northern Ireland. From the point of view of disease, the stretch of water between Northern Ireland and GB, together with the great vigilance of all involved, has been of considerable economic advantage to Northern Ireland over the years.

Edwin Conn, then chief veterinary officer, recalls the outbreak of FMD in England:

> It was during the late autumn of 1967 when I got an alert about foot and mouth

from a chief veterinary officer friend in Britain. We have always had a close relationship with our colleagues in all parts of the UK and the Irish Republic. I called my senior staff together, briefed them accordingly on the possibility of major problems and we decided there were certain things we needed to do.

The slaughter policy applied during the outbreak of foot and mouth disease in England, 1967

DARD

Close connections with his government counterparts enabled Edwin Conn to keep his finger on the pulse during the crisis in 1967: 'You have to recognise the speed with which FMD can spread. In the late 1960s' outbreak we had tremendous co-operation from everyone.'

The Ministry of Agriculture continued its general policy of testing all herds every two years for tuberculosis. Any reactors were purchased by the Ministry, and slaughtered, and compensation was paid. The incidence of the disease was less than 0.1 per cent and, from November 1960, Northern Ireland was declared an attested area for the purposes of tuberculosis in cattle. Throughout the 1960s the disease was virtually non-existent in Northern Ireland. The eradication programme for brucellosis, which started in 1963, resulted in over 98 per cent of herds being certified free of the disease by the end of the decade. Similarly, the Ministry was anxious to eliminate warble fly infestation from cattle. A two-year programme began in 1966–67 to dress all cattle, except calves or exempted animals, with a systemic insecticide to control warble fly infestation, with a view to eliminating this insect pest. The proportion of cattle affected by warble fly was reduced to a negligible level following completion of the programme in 1968.

Examining milk samples for mastitis-causing bacteria, a problem which remained in many dairy herds

DARD

There were concerns, however, about the levels of other disease among farm animals. Ailments and infection among intensively housed livestock had risen appreciably, at inestimable annual expense to the industry. A comprehensive milk-testing programme was operated by the Ministry to identify herds with a high incidence of mastitis. Liver-fluke infestation and other parasitic conditions continued to be a drain on the livestock industry. Fears were expressed that certain organisms, including typhoid and salmonella in poultry, could develop resistance to antibiotics. There were also claims that administering antibiotics to animals had already harmed human health; Dr David Luke had expressed these fears in the 1950s.

Concerns about safety were also beginning to emerge in relation to standards of husbandry. In the UK there were warnings of health risks from 'factory farming'. Anne Coghill, chairperson of the Farm and Food Society, warned farmers that they couldn't expect the nation's support if their fight for a better price review was based on higher productivity, lower cost and inferior produce. There was, she asserted, 'a mounting pile of evidence' which proved that factory farming methods throughout the 1960s held many health hazards for the consumer. 'What a large thinking proportion of the public would like to hear,' she said, 'is that farmers are fighting for better farming and not merely their own betterment.'

Pollution

Productivity was increased through the application of weedkillers to grass and cereal crops.

DARD

The acreage of grass silage increased from 94,000 in 1963 to 147,000 in 1970, when production exceeded 1.5 million tons. This represented an increase of about 70 per cent in the amount of silage made over the last three years of the 1960s. One of the associated problems that became apparent was the general

increase in farm effluent, many cases of pollution arising from silage-making. It was the farmers' responsibility to dispose of effluent from their farmsteads in such a manner that it did not create a nuisance, pollute water supplies or contaminate watercourses. Farmers were being forced to ensure that farm effluent did not cause such problems. There were, however, some offenders, and incidents of fish kills did occur. When bacteria break down the organic matter in farm effluent, dissolved oxygen in the water is removed and fish are unable to survive. The scientific term is biochemical oxygen demand (BOD), a measure of the amount of dissolved oxygen used up when polluting material is broken down. Silage effluent has a very high BOD. Silage effluent was to be a continuing problem until the end of the century.

Another issue that raised its head in the mid-1960s was the question of toxic chemicals. Previous years had seen the increasing use of chemicals such as weedkillers and pesticides, resulting in great improvements in production. There was, however, considerable comment from outside the agricultural industry about the effect of pesticides on the 'balance of nature'. Most farmers exercised great discrimination when using these toxic chemicals but some misuse probably did take place.

Safety on Farms

Farm safety was a key issue in the 1960s, and attention became focused on the number of farm fatalities. In 1968, for example, twenty people lost their lives on farms, a number of them in accidents caused by overturning tractors or related to tractor-handling. It was described as 'a black year'. In an attempt to highlight dangers on the farm, the Ministry of Agriculture mounted a tough safety campaign in 1968–69. A booklet entitled *Keep Farming Safe* was circulated to some forty thousand farmers. Together with their employees, they were alerted to the dangers of unguarded power take-off shafts and to the hazards of uneven weight distribution on tractor wheels. Exhibitions and demonstrations on farm safety were arranged. New legislation was promoted in 1968 on the question of safety cabs and frames for agricultural tractors. Cabs afforded tractor operators greater protection. Machinery was becoming a great boon for farming but, unfortunately, it also brought dangers.

By the end of the 1960s, slurry and effluent tanks were playing an increasingly important role on farms. During storage, however, slurry starts to break down and produce gases, some of which are poisonous or even explosive. When slurry is disturbed, these gases can be released, creating a dangerous situation. The danger from slurry tanks was highlighted as part of the farm-safety campaign. Farmers were also strongly advised on how to prevent farm fires, guard against electrical faults and handle herbicides, chemicals and sprays with

An accident on a farm. In such a situation, a safety cab could prevent the driver from sustaining serious injuries.

DARD

greater care. The Minister of Agriculture in June 1968, Major the Right Honourable J.D. Chichester-Clark, hoped 'that the sum total of all our [the Ministry's] efforts will be a substantial reduction in farm accidents'.

Financial Concerns

In spite of the numerous grants and subsidies, farmers in Northern Ireland were still unhappy with their financial situation, especially in comparison with the rest of the UK. Their problems were brought to the attention of newly appointed Minister of Agriculture, Cledwyn Hughes, by a delegation of Ulster MPs led by Captain L.P.S. Orr and the Marquis of Hamilton. They told him that Northern Ireland farmers were 22 per cent worse off than their counterparts in GB, a figure that didn't even take into account the extra burden of £4.25 million they had to bear through devaluation. They called for an increase in the Remoteness Grant (under which Northern Ireland farmers received £1.75 million per annum from 1966–67 to 1970–71). The grant was paid to help alleviate the disadvantage arising from Northern Ireland's distance from its main markets in GB. It was originally set in 1957 at a level of £1 million per year for five years.

The delegation emphasised that farmers in Northern Ireland had experienced lower incomes. Concern was also expressed about the future of the egg industry which at the time (1966–67) was worth £17.6 million per year, and which represented 15 per cent of overall agricultural production in Northern Ireland. While the annual review in 1967 brought £2.1 million in increased prices and guarantees to farmers in Northern Ireland, dairy farmers expressed concerns that an increase of 1.5d a gallon for milk was inadequate compensation for the work and expenditure involved. Some farmers did feel that the government had made a reasonable effort to meet the needs of the industry, but others believed the increases obtained for products and livestock did not compensate for the setbacks suffered over the previous five years. The annual farm price review was of paramount importance to everyone involved in Northern Ireland agriculture in the 1960s. Sometimes the outcome was received with gratitude, while on other occasions it was condemned for falling far short of farmers' expectations.

When debating the economic development of Northern Ireland, the Minister of Agriculture, Harry West, took the opportunity, in February 1965, to explain the role that agriculture might play in this. He stated that increasing farm income should be the objective, with the government

> … doing everything possible to lower costs of production and thereby increase the margin of profit from existing products; improving marketing arrangements when we can; and continuing to make careers outside the agricultural industry more readily available and more attractive, so as to encourage many of our smaller farmers and their families who cannot easily make a living on the land to leave the agricultural industry and take up alternative employment. This, in turn, will enable other farmers to purchase their farms and thus bring about the amalgamation of small farms to form larger units.

Harry West, a thousand-mile-a-week motorist, in the days before the creation of the M1 and its sister highway, the M2, was ever at the beck and call of UFU members. They did not forget him either when he got married in August 1956. They presented the man, known as the earliest rising MP, with a new £750 motorcar. It was indeed an appropriate present for it was said that he had 'worn a car out' during his year of UFU presidency. 'Indeed I did,' he said at his Ballinamallard farm. 'There was a lot to do and I'm not so sure now in hindsight if I would do the same again – looking after two or three jobs at the one time. Then again, who knows? I probably would do it all over again.' Often he was at UFU headquarters in Belfast before 9 a.m. Not bad going for someone who had just travelled one hundred miles. For good measure, he often covered the journey three times a week and fitted in a trip to London as well, on union business. 'I remember when in Stormont I tried hard for all farmers and tried particularly to get more money for those involved in dairy farming. I suppose I can say I did enjoy being Minister of Agriculture at the time.' He remembers too the setting up of such entities as Moy Park as a farmers' co-operative, and characters on the scene like the late Eddie Swain, a former chairman of the now defunct PMB, George Cathcart and George Ervine.

Would he go back into farming again given the opportunity?

Yes, I would, but it is a different industry now to the one I knew and worked in. There have been dramatic changes over the years and there will be many others to come. I liked the land. It was in my blood. You need to have a deep love of the land if you want to be a farmer. I think that is what's wrong with some people today. They want to take everything they can from the land but what do they put back into it?

Harry West was keen on farmers having a say in the running of their industry and in the marketing of their products. 'I think that scene has changed dramatically and the farmer tends to be the servant of the supermarkets.'

Ulster Farm in the 1960s

During the 1960s, throughput at the Glenavy rendering plant increased dramatically from 200 to 500 tons maximum. It was a period of expansion and consolidation against continuing financial difficulties. There were two cookers at the start, and this had increased to seven by the end of the 1960s. The number of extractors had also increased from one to three over the same period. New milling plant was also installed during the 1960s. At this stage, including lorry drivers, Ulster Farm employed 120 people.

The material supply was irregular and seasonal, with a glut of animals in the spring, causing processing inefficiencies. Offal material was, however, secured under contract from the pig industry and meat plants, which helped Ulster Farm to overcome its financial problems during the 1960s. With the development of meat plants in Northern Ireland and a higher proportion of beef animals being killed prior to export – as opposed to being exported live – a greater supply of raw material became available for Ulster Farm, as the co-operative expanded its services to these new plants.

Conclusion

The pattern of farming was changing as the 1960s moved on. Previously, there had been many small-scale holdings, involved primarily in mixed farming. Because of the difficulty in making a reasonable income from these units, farming started to become more specialised. The number of activities or enterprises on each farm became less than in previous years, and the farms became slightly larger. Improved farming methods required further investment of capital. As a result, productivity improved throughout the 1960s. There was also renewed thinking in relation to the export of cattle from Northern Ireland to GB. Development of a number of privately owned meat plants and abattoirs resulted in more added value accruing to Northern Ireland when carcases, rather than live animals, were exported.

In general, this was a period of prosperity and growth for farmers in Northern Ireland. Government grants and subsidies enabled them to continue the process of modernisation and to make further farm improvements. Agricultural development was fuelled by government investment through an improved advisory service and by the expansion of agricultural training opportunities. In spite of these advances, financial concerns were beginning to surface, and it is worth mentioning two developments that occurred towards the end of the decade. They had limited impact at the time but would have great significance for the 1970s. The first development was the devaluation of the pound sterling in November 1967, under the Wilson government. This laid the basis for the inflationary spiral that took off in 1973. The second development was the drop in beef prices in the mid-1960s and the international difficulties in the dairy market. These problems had little effect on agriculture in Northern Ireland at the time because of the protection afforded by the 1947 Agriculture Act. The devastating effects of these crises would, however, impact severely on farmers in the 1970s.

Speaking in May 1969, the Minister of Agriculture, the Right Honourable Phelim R.H. O'Neill, said that two major trends would increasingly affect agricultural policies:

> One is the increasing pressure on profit margins resulting in a lowering of profits per unit of production on farms. The other is the growing tendency of housewives and retailers to prefer foods which have been processed and packaged in more sophisticated ways.

These trends were to become very evident in the decades ahead.

1970s

Introduction

The key event for Northern Ireland, and indeed UK, agriculture in the 1970s was entry to the European Economic Community (EEC) in 1973 and the acceptance of the principles of the Common Agricultural Policy (CAP). Prior to entry, there had been great anticipation that prices for farm produce would improve. However, EEC entry coincided with generally difficult economic conditions and with depressed employment levels. This led to reduced confidence within the farming community, which had already suffered major falls in the price of beef, milk and other farm produce.

As in previous decades, farm improvements and continuing modernisation advanced in the 1970s. Additional bulk feed-bins appeared amongst farm buildings as more farms became equipped to handle and store feeding stuffs in bulk. The number of farms with milking parlours and bulk milk-tanks continued to increase. By 1980, capital investment in tractors, vehicles and equipment was estimated to be running at £33 million a year, in comparison with some £6 million at the start of the decade.

Cattle numbers on farms stood at 1.4 million in 1981, a marginal increase over 1971. There was little difference in total cow numbers between 1971 and 1981, although the figure had risen to 580,000 in 1974. About 90 per cent of farmers' incomes came from the sale of livestock and livestock products. By 1981, the area devoted to crops stood at about 75,000 hectares, a drop from 97,000 in 1971, and only one fifth of the area recorded in 1847 when farm census figures first became available. The reduction in tillage reflected a decline in the area of all crops other than barley.

As they moved into the 1970s, farmers had little inkling of the beef crisis that lay ahead in 1974. Weather-wise, it was a severe period, with 1974 being a very wet year, and 1975 and 1976 being amongst the driest on record. The year 1979 was also one of physical difficulties, described by the Department of Agriculture in the following terms:

> the long and abnormally severe winter, the late spring, the subsequent problems in securing winter fodder supplies and in saving the harvest; the lower prices for fat pigs, summer fattened cattle, suckled calves, and store and fat lambs compared with 1978 and, not least, the continuing upward surge in farm costs.

Quite a depressing litany! There seemed little alternative but to strive for greater efficiency in every aspect of production as the scope for improving product prices was very limited under the CAP. Nevertheless, Northern Ireland farming had a total gross output of almost £612 million in 1981 from its farm businesses. The agricultural export trade was also booming and, by 1981, exceeded £420 million a year, excluding dairy products.

Pressures on Farming Structure and Businesses

Throughout the 1970s, there was sustained pressure on the structure of farming and on individual farm businesses. Economic strains drove more people off the land, and the gradual trend towards larger farms remained. Both the number of bigger farms and their share of total output increased during the 1970s. By 1979, the average area farmed in Northern Ireland was almost 26 hectares of crops and grass, compared with 48 hectares for the UK and only 18 hectares for farms in the EEC. In 1979, it was estimated that there were some 31,000 farm businesses considered to be significant to the industry. These farms were producing more than 85 per cent of total agricultural output. Livestock production remained of major importance and, in 1981, 75 per cent of the full-time farm businesses (about 19,400) were engaged mainly in either dairying or beef and sheep production. In addition, there were about 970 specialist pig and poultry farms and some 950 specialist cropping units. The remaining 1,600 full-time farms were 'mixed' and did not have any predominant single enterprises.

Two schemes to improve farm structure were introduced in September 1973, replacing earlier schemes. They were aimed at increasing the size of farms and improving their layout. The 1973 Farm Amalgamation Scheme encouraged the voluntary amalgamation of uncommercial farms with other land, to create commercial units, or intermediate units as a step towards becoming commercial. Grants were available to help amalgamators with their new units and to encourage outgoers to leave the land. The scheme closed for new applications at the end of 1976. The 1973 Payment to Outgoers Scheme encouraged farmers occupying uncommercial farms to relinquish them for amalgamation with other land. Acreage payments were made to those who carried out approved amalgamations, absorbing land on which an outgoer's payment was made. Outgoers had the opportunity to submit applications under these or similar schemes throughout most of the 1970s.

Robin Morrow, who went on to become president of the UFU and chairman of the Northern Ireland MMB before it was deregulated in 1995, farmed 150 acres at Ballyhanwood outside Belfast. His prediction at the time was that the number of farms would be drastically reduced to as few as 15,000. 'We would love to be able to farm as our fathers farmed but economic pressures have forced changes on us,' he stated. The drop in the number of farm workers and in the number of working owners, which had begun in the 1960s, was maintained during the 1970s.

By the 1970s, farmers in Northern Ireland had not only had to absorb a range of new farming techniques and methods, but they had also had to learn to cope with a number of new responsibilities and roles. Michael Quinn, development officer with the highly successful Richardsons Fertilizers in the early part of the decade, summed up the demands and pressures on farmers:

An example of an average farm might well be a sixty-acre holding carrying 40 dairy cows and 20 younger cattle, with some £35,000 invested in land, buildings, stock and machinery. Such a substantial investment of capital meant that farmers, like owners of any business, needed to have good management skills.

Michael Quinn strongly advocated good record-keeping, as well as the need to budget carefully while keeping a sharp eye on ever-rising costs. His message to farmers was:

> The organisation of labour and machinery, and the substitution of manpower by mechanical means requires careful planning. Like all management, modern farm management involves a constant cycle of recording, analysis, planning, action, measuring and review. But unlike other businesses, the farmer himself usually plays the role of manager, accountant, financier, machinery and crop and animal production expert, as well as being a large part of the farm's labour force.

The extent to which farming had changed can be gauged by looking at the McGuckian brothers, Alastair and Paddy, who were later to win worldwide orders and much acclaim with their unique Masstock livestock systems which brought maximum ventilation and improved conditions to housed livestock. They farmed seven hundred acres at Masserene in County Antrim and had thousands of cattle on their farm, but not one animal in the fields. The ruthless efficiency of intensive farming meant that the animals were kept indoors under a system of husbandry involving carefully controlled conditions. The McGuckian brothers had built up a capacity of 2,500 cattle, which turned over up to twice a year, and the whole beef enterprise was run by only seven men who were highly trained operatives specialising in beef production. The pressures to intensify the farming industry proceeded relentlessly during the 1970s.

Discontent on the Farms

Structural and business-management pressures were not the only problems facing farmers at that time. They had concerns about their incomes too. The debate about entry to the EEC was taking place against the background of an agricultural industry undergoing continuous evolution. Given the various pressures, it is perhaps not surprising that the 1970s opened with a discontented farming community. In February 1970, with the annual review looming, farmers blocked streets in Tyrone with a cavalcade of tractors, in an attempt to bring their concerns to public attention. 'Bare bones Price Review won't be tolerated' said one of the placards. Lisbellaw farmer Eric Donaldson spelt out what it was like to work a thirty-five-acre holding. 'You haven't got the same freedom as anyone in a job,' he said. An average day began for him at 7 a.m. when he got up to milk the cows, and ended some twelve hours later. Holidays were few – maybe a rare day trip to Bundoran which had to be cut short to enable him to be back in time for evening milking. 'You can get a lot better money in a factory. I heard of a young fellow of twenty-three who is getting £24 a week in a factory. You would work a good bit on a farm before you'd get that,' he told a *Belfast Telegraph* feature writer at the time.

In the early 1970s, the government introduced a development programme for the period 1970 to 1975, in an effort to jump-start the economy as a whole. One part of the plan was intended to compensate Northern Ireland livestock producers for the disadvantage that they encountered through having to buy higher-costing feeding stuffs. Approximately £0.5 million per annum was paid to farmers producing pigs, eggs and poultry meat, milk and beef cattle. The government tried to encourage greater production from grassland by reducing the cost of nitrogenous fertiliser. The amount spent under the programme was in proportion to the amount of purchased feeding stuff used by each particular enterprise. The government paid out a further £9 million over a three-year period in assistance to egg producers, pig farmers and grassland farmers. This was aimed at helping to increase stocking densities through land reclamation and grassland improvements.

The 1974 Farming Crisis

In 1974, the Northern Ireland beef industry was facing its worst ever crisis. There was too much beef around and not enough people to buy it. Overproduction had caused an international beef surplus and a sharp fall in prices. There was widespread disillusionment, followed by a loss of farming confidence. Northern Ireland, a traditional beef-exporting area, was severely affected, and bore the brunt of the crisis. James T. O'Brien of the LMC said that the root cause of the crisis 'was allowing in too much imported beef. Cattle were coming on the market at 25p per pound or around £130 per animal but that was a price well below farmers' costs.' It was estimated that farmers were losing around £40 on every animal. The government did attempt to help them by means of a slaughter premium of £9 per animal, but this fell far short of the amount of money being lost.

The situation deteriorated rapidly, and farmers took to the streets in 1974, actually threatening to give away their cattle in protest. There were warnings of a general strike by the twenty-eight thousand UFU members, an action that could result in the withdrawal of milk supplies. In Dungannon, more than four hundred farmers marched on the Crown Buildings to sign on as unemployed and to draw attention to their perilous position. Farmers' despair reached its height in mid-September 1974 when thousands attended a mass protest. Cavalcades of tractors, Land Rovers, and farm machinery paraded through provincial town centres. Protesters travelled from outlying areas to centres such as Coleraine, Portadown and Ballyclare to focus public attention on the crisis. Around thiry five-thousand leaflets entitled 'Your food is at risk' were handed out to passers-by. Townspeople were warned by the farmers that if 'we go broke – you go hungry'! In Limavady, one farmer marked the occasion in macabre fashion by carrying the body of a slaughtered cow in a trailer at the front of the cavalcade. Vehicles carried slogans such as 'No sugar today, meat and eggs next'. In Portadown, bowler-hatted farmers in mourning suits formed a funeral procession with a mock coffin draped in black. The protest in Newtownards saw practically every kind of farm machinery out on the roads, and in Belfast 130 vehicles took over the Carryduff roundabout.

Protesting farmers at Carryduff block the main road in south Belfast.

Belfast Telegraph

In the wake of the protest, it became known that the number of cattle that would normally have been artificially inseminated had dropped by a staggering 30,000. It was reckoned that beef prices had been slashed by 50 per cent in a twelve-month period. Farmers, unable to cope with the drop in prices and faced with rocketing fodder prices, were giving up farming and selling off their animals quickly. The reduction in herds caused a drop in output from the beef sector, and forecasters predicted that production would be unlikely to catch up on consumption before the early 1980s. Land prices fell sharply in the wake of the 1974 crisis, mainly because of the many farmers who were forced to sell. In the House of Lords, Lord Brookeborough called on the government to restore farmers' confidence by taking steps to improve the economic situation.

When the UK joined the EEC, it was not anticipated that, within months, events on the world cereal market would play havoc with farm profits. There was an acute world shortage of cereals and protein raw materials for animal feed in 1973. Russian purchases of most of the American reserve grain stocks led to rocketing prices of feed grains worldwide. Added to this was the oil crisis of October 1973, which raised farmers' production costs even higher. The escalation of cattle numbers to a record level in 1973–74 had also been an important factor. Poor demand for the larger number of cattle being sold in late 1974, together with inadequate price-support measures, resulted in depressed prices which severely dented the confidence of beef producers.

The situation had been exacerbated by the general international economic climate. Inflation had risen sharply because of constraints on industrial manufacturing capacity and depreciation in the value of sterling and the dollar. This, in turn, forced the major oil-producing countries to increase the price of oil significantly, resulting in a steep rise in unemployment levels throughout the UK. The 'green pound' was a representative rate at which support prices were translated into national currencies. Its devaluation in the Republic of Ireland had severe consequences in Northern Ireland and cast a doubt over the future of bacon factories and meat plants which were being aided by the Meat

Farmers protesting at Stormont over the crisis in farm incomes

Belfast Telegraph

Industry Employment Scheme. The purpose of this scheme was to offset the difference between the Irish Republic and the UK 'green pounds', help the slaughtering and processing industry in Northern Ireland to obtain supplies of cattle and pigs, and thus maintain employment.

Farming in Northern Ireland also suffered further setbacks during 1974. The Ulster Workers' Council strike in May 1974 brought life almost to a standstill and brought down the newly formed Northern Ireland Executive. There was temporary disruption to electricity supplies to farms and industrial plants, causing difficulty in the production and processing of some agricultural products. The Royal Ulster Agricultural Society (RUAS) May show was postponed until June, and the strike cost the industry in excess of an estimated £5 million. Many grain farmers also experienced their worst harvest since the 1950s because of inclement weather conditions.

The severity of the farming depression was, however, not as great in Northern Ireland as in other parts of the world, because of the effect of the transition of UK agriculture into the EEC. The drop in the value of sterling was compensated by 'green' currencies. Ironically, it was in the middle of the beef crisis that Ulster Farmers' Investments Limited (UFIL) came into being. The LMC had begun an investigation into the involvement of producers in slaughtering and marketing of meat. UFIL, a farmers' co-operative, brought a number of beef producers together in a common cause. They raised over £1 million, thanks to financial support from the UFU and from the Department of Agriculture. This enabled the co-operative to purchase a 100 per cent holding in Moy Meats in 1976, becoming the first producer co-operative in Northern Ireland to own and run a meat plant. Moy Meats was operated by UFIL for almost twenty years and served the industry well by providing necessary competition for the purchase of cattle by a farmer-owned co-operative.

Entry to the European Economic Community

The UK first applied for EEC membership as far back as 1961, an application that was vetoed in 1963. A further application was made in 1967 but this, too, was vetoed. In 1969, the EEC indicated that it was ready to resume negotiations on UK entry, and these began in July 1970. At that stage, the EEC had a membership of six countries – France, Italy, the Federal Republic of Germany,

Belgium, the Netherlands and Luxembourg – 'the Six'. Their combined population in 1968 was 186 million, making the EEC one of the largest trading units in the world.

Although the EEC was heavily industrialised, agriculture was still an important employer of labour in all member states, so much so that a common policy was adopted for agriculture – the CAP. It was basically a 'managed market system'. Farmers were expected to get most of their return from the market, making it a 'high-price' system.

James O'Brien recalls that, during George Cathcart's term as UFU president, there were serious discussions about membership of the EEC. 'I never had any doubts or misgivings about the wisdom of membership,' he states, explaining:

> In general most men of my generation who had served in the war and experienced the unnecessary carnage, destruction and destitution, hoped for the vision of harmony and mutual understanding that could ensue from an engendered community spirit. But leaving aside such idealism and thinking in wider industrial terms, I saw advantages. My experiences at the annual reviews had led me to think that a change to a system of support for agriculture, based on import controls and intervention buying to support the price level was preferable to the system of target prices and deficiency payments.

Considerable discussion took place on issues such as the annual review of agriculture; the EEC's arrangements for milk, pig meat and eggs; the provision of assistance to hill farmers; and the position of New Zealand dairy products. Eventually, the Right Honourable Anthony Barber, Chancellor of the Exchequer and minister in charge of the negotiation at the time, stated that, as a member of an enlarged community, the UK would be prepared to accept the CAP. The terms for UK entry to the EEC were published in July 1971, although the transition period did not begin until January 1973. With the addition of the Republic of Ireland, Denmark and the UK, the enlarged community of 'the Six' became 'the Nine'.

In a memorandum in 1968, Commission Vice-President Sicco Mansholt, chief architect of the CAP, set out his views on the reform of EEC agriculture. Too much money had been spent 'on bolstering prices of surplus products', he asserted. He wanted a different price policy, aimed at restoring a more normal relationship between market and price trends. He advocated bringing farms up to a viable size to enable farmers to live as comfortably as everyone else. 'Financing farms with five cows,' said Mr Mansholt, 'is tantamount to financing chronic destitution.' The plan's basic premise was that the EEC's farming population of 10.6 million should be reduced by five million. Mansholt clearly wanted change. He summed up the position reached in 1970:

> The Commission's price policy, based on consensus politics, rather than economics, has taken us to the end of the road, with structural surpluses costing astronomical amounts. Dairies churn out subsidised butter regardless of market needs; no one worries about packing the stuff, because no matter whether it is bought, stored, sold cheaply or destroyed, the producer gets the guaranteed price.

Against this background, Britain joined the EEC, the Common Market, on 1 January 1973. The CAP began to replace the system of price guarantees that had been operating in the UK until that time. There was a five-year transitional accessional period until 1978, during which time UK prices were brought up to EEC levels. Meanwhile, farmers and others were deluged with all sorts of terms in relation to the commodity prices within the CAP – indicator prices, intervention prices, threshold prices, sluicegate prices, guide prices. The list seemed endless. Suffice to say that the objective of achieving comparable prices by 1978 was achieved throughout the EEC. The big event in the agricultural sphere then was the UK's formal accession as a full member in 1978. This meant, of course, that the industry was then subject to all the ramifications and responsibilities of full EEC membership.

As the prices of most agricultural products were higher in the EEC, farmers in Northern Ireland viewed the new situation with optimism, many anticipating that it would bring better prices. Their confidence came, largely, from the safety net of 'intervention buying' that was introduced in 1974. This mechanism allowed the EEC to buy in products from member states when there was a glut in the market, or to prevent such a glut developing. The products were then stored until prices were higher, or were sold off at reduced prices to countries like Russia. In the year following accession to the EEC, Northern Ireland enjoyed the benefits of intervention buying, particularly in the beef sector. Farmers were producing more than they could sell on the open market but they still achieved an acceptable price. Intervention buying by the EEC created a foundation for the market and helped to keep beef production afloat in Northern Ireland.

Government Assistance Schemes

In many ways, farmers' confidence in the EEC was rewarded. Entry into the EEC brought with it a plethora of grants and subsidies to facilitate farm modernisation and improvement. Two major new EEC-sponsored schemes came into operation in January 1974. They were the Farm and Horticulture Development Scheme (FHDS) and the Farm Capital Grant Scheme (FCGS), 1973. A third scheme, the Horticulture Capital Grant Scheme, also commenced in January 1974, providing assistance for certain horticultural investments. They were all implemented under a new EEC farm modernisation directive.

The FHDS was set up to provide aid for farmers who wanted to modernise their holdings. It was open to farmers whose income per labour unit was less than the national average of workers in non-agricultural occupations (£2,070 at the time in Northern Ireland) and who submitted a plan to reach this level within six years. Capital grants were paid on the investment needed to carry through the plan; a grant was given towards the cost of keeping farm accounts; and, under certain conditions, a special guidance premium was payable when more than half of the farm's income was derived from the breeding of cattle or sheep for meat. The general rate of grant was 25 per cent, but field drainage qualified for 60 per cent.

The FHDS attracted more than three thousand applicants who planned to invest around £80 million in their businesses. More than 50 per cent of applications came from farmers in the dairy sector, and they received grant-aid of £27 million. There were warnings, however, that too much milk was being produced within the EEC, but this did not halt UK plans for increased milk production. A proposed five-point EEC action plan to deal with dairy surpluses left many milk producers fearful. The plan envisaged the introduction of non-marketing premiums, suspension of aids to the dairy sector, and a levy on milk producers. The plan would eventually result in the introduction of milk quotas in 1984.

The FCGS was particularly important, providing grant-aid on a wide range of capital improvement works. Farm buildings, yards, sheep and cattle grids and livestock handling facilities were all eligible. At the outset, the standard rate of grant covered 20 per cent of the approved cost, but the rate for building cattle accommodation, dairy and parlour buildings and grass silos was higher, at 30 per cent. For field drainage the rate was 50 per cent, with a 70 per cent rate when the work would benefit an agricultural business in the Less Favoured Areas (LFA), in which the income per labour unit was less than the average earnings of full-time workers outside farming. Farmers were quick to take advantage of these grant opportunities, although the rates of grant were

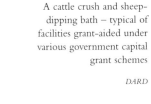

A cattle crush and sheep-dipping bath – typical of facilities grant-aided under various government capital grant schemes

DARD

reduced later in the 1970s. In the year ending March 1979, some 21,500 applications for approval were received under the scheme, and grant paid out on completed works amounted to almost £11 million in that year.

As part of the annual review settlement in 1974, the fertiliser and lime subsidies, which had been payable for many years, were phased out. It was also decided that because of EEC membership the Remoteness Grant would no longer be appropriate and it came to an end in March 1977. For the last three-year period, the grant had been approximately £1.9 million per annum.

Hill Farming

The new schemes did not provide any special rates of assistance for hill farmers, pending the adoption of the relevant EEC directive. However, under the 'Directive on mountain and hill farming and farming in certain less favoured areas', certain areas were designated as 'Less Favoured'. The LFAs were defined by the type of land and physical conditions such as topography and altitude. They were essentially those areas in which Hill Livestock Compensatory Allowances (HLCAs) were paid on eligible stock.

In Northern Ireland, they comprised the 'hill' areas, together with certain additional poor areas of heavy wet land. The LFAs were particularly important for beef and sheep production. In area, they covered about 45 per cent of the agricultural land in Northern Ireland in 1978. At a later stage, following representations to the European Commission in 1982, the LFAs were extended to around 70 per cent of Northern Ireland land.

HLCAs were implemented from January 1976 as part of the EEC directive designed to help farmers in the LFAs. They replaced the hill cow and hill sheep subsidies that had been paid previously. Farms in these geographical areas could apply for special grants and funding from the EEC to help overcome problems, such as soil quality, which resulted from their location. The allowance was paid to owners of regular breeding herds of cows kept for breeding and rearing calves and to the owners of breeding ewes maintained on eligible land. To qualify, claimants had to occupy at least three hectares of land and, in general, they were required to sign an undertaking to continue farming for the next five years. Payment was subject to an overall financial limitation based on the area of eligible land available for cattle and sheep, with payment on sheep limited to six ewes per hectare. In 1977, some 13,200 applications were received, and £4.4 million and £1.1 million was paid out for cattle and sheep respectively.

Escalating Land Prices

After joining the EEC in 1973, price levels of UK farm produce were gradually aligned with those of the other members. It was inevitable that the price of farming land would also rise but it did so, disproportionately, between 1976

Table 4: Average Price of Land Sold in Northern Ireland: 1973 to 1978

| Year | In Current Terms | | | In Real Terms* | | |
	£/hectare	£/acre	Index	£/hectare	£/acre	Index
1973	825	334	100	825	334	100
1974	1,000	405	121	862	349	104
1975	1,150	465	139	798	323	96
1976	1,375	556	167	818	331	99
1977	1,850	749	224	953	386	115
1978	2,620	1,060	317	1,241	502	150

*Adjusted for inflation
Source: Ministry of Agriculture: Agriculture in Northern Ireland, December 1979

and 1978. If the prices paid for land were to be based on earning capacity alone, one could not understand how the dangerously high prices paid during 1978 could be justified. Even in real terms (adjusted for inflation), the price of land had increased by 50 per cent in six years. Table 4 shows the average price per hectare or per acre of agricultural land sold in Northern Ireland from 1973 to 1978.

The high land prices of 1978 caused considerable discussion at farmers' meetings. The view often expressed was that many purchasers would 'never see their money'. Land prices were determined in the same way as the price of other commodities, being dependent on demand in relation to supply. Farmers wanted to increase the size of their farms for many reasons. These included the expansion of their businesses to increase their earning capacity, to the purchase of adjoining land that came on the market, or to directing spare cash towards a secure inflation-proof investment. However, on occasions, farmers were known to expand their farms irrespective of cost. And problems can arise when land is purchased with borrowed funds and the earning capacity is not high enough to repay the capital and interest charges (because loans always have to be repaid). This certainly seemed to occur in 1978.

Decimalisation and Metrication

The changeover to decimal currency took place on 'D Day', 15 February 1971, when farmers and the general public said goodbye to £sd and began using decimal £ and p. Everyone was warned that the banks would close on 11 and 12 February to convert their books and accounts, and would re-open on Monday, 15 February, when they would account wholly in decimal. Farmers, too, had to change their records and accounts to the decimal system. There were many implications for farmers – for example, in the payment of wages and the calculation of income tax and graduated contributions. The Ministry's payment cheques, an important consideration, were also paid in decimal currency from 'D Day'.

In 1972, James Prior, then Minister of Agriculture in London, announced that, following consultation with farming organisations, the government would also support a metrication programme for agriculture, centred on the farming year 1975–76. The change was considered necessary to keep in step with international trade. Farmers' customers, suppliers and competitors were all committed to going metric. This meant that the annual review and the agricultural census had to take place in metric terms for the first time in 1976. The dairy sector of the industry was one of the first to go metric – milk was collected from farms in litres instead of gallons in October 1976. The 1000-gallon per lactation cow became the 4,500-litre cow. The 10-cwt bullock became the 500-kilogram bullock. Terms such as pounds, gallons, yards and acres were to be replaced (in theory) with kilograms, litres, metres and hectares. Pigs were to be housed at a temperature of 20°C rather than 68°F. Fertiliser would be applied at 75 kg per hectare instead of 60 units per acre. In practice, not all farmers found it easy to adjust to using the metric system. Even in the year 2002, there are still many farmers who, in conversation, refer regularly to the old system.

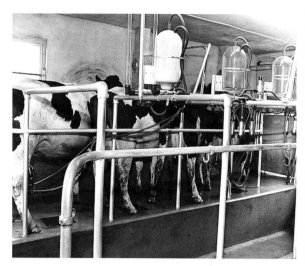

Developments in milking parlours meant that more cows could be looked after by fewer people.

DARD

Production Trends

The production of fat cattle and milk production were still the most important livestock sectors throughout the 1970s, accounting for approximately 35 and 24 per cent respectively of total gross output in 1980. From 1973, the total cattle population had stabilised at just over 1.5 million, having expanded throughout the 1950s, 1960s and early 1970s. Between 1967 and 1973, in particular, the total number of cattle in Northern Ireland rose spectacularly by some 25 per cent to a figure of 1.54 million, probably in response to the expansion exhortations. Higher cattle numbers meant that more land had to be brought under grass, reducing the acreage of cereals and potatoes by almost 40 per cent.

Dairy herds continued to increase at a moderate rate, with progressive dairy farmers making more use of loose housing as lying-in quarters. Most animals in the larger herds were now housed in winter in cubicles or kennels, leading to cleaner conditions for the cows. Sophisticated milking parlours, both herring-bone and rotary type, were being equipped with automatic feed dispensers. Over the course of the 1970s, total dairy cows increased from 215,000 in 1971 to 270,000 in 1981. The total output of milk produced increased during the 1970s, and by 1981 had reached 1,190 million litres, valued at £149 million. The guaranteed price was increased at successive annual reviews, based on the standard quantity of milk that was used for liquid consumption. However, because the amount of milk produced was considerably above the standard quantity, the price received by producers was well below that of their GB counterparts. In 1976, the estimated average yield per dairy cow exceeded 4,000 litres for the first time, reaching 4,420 litres at an average producer price of 12.57p per litre in 1981.

Regarding beef production in the 1970s, the imported beef breeds from the Continent were making a useful contribution to better quality calves. Feeding systems also developed, with more use being made of high-dry-matter silage. However, the beef slump of 1974 resulted in a major decline in cow numbers, with the loss of 48,000 beef cows in the following two years, and a further 86,000 by 1981. The very poor price for beef in 1974 meant that some drastic measure was required to restore the confidence of beef producers. This came in the form of the Variable Beef Premium Scheme, beginning from March 1975. This provided a means of maintaining producer returns for finished cattle in line with target prices. The Minister of Agriculture, Rt Hon. Fred Peart, when referring to the 1975–76 annual review, said that 'the most important decision of the Council was to make radical changes in the Community regime for beef'. The Minister continued: 'These premiums will enable the government to assure producers of a fair return.'

Marketing of cattle in the early 1970s still showed that too many were being sold in autumn and winter, rather than in spring and early summer. The uneven pattern was causing problems for the meat plants. The overhead costs per animal were too high and the plants were having difficulty in meeting their customers' demands. As a result, farmers were encouraged to market more of their cattle in the February to June period. Further encouragement was given to slaughtering fat cattle in Northern Ireland as opposed to exporting them live, a trend that had begun in the late 1960s. In addition to creating employment in Northern Ireland, this also helped to counter growing public concern about the possible stress caused to live animals in transit to GB. In any event, the shipment of live cattle was subject to rising and fluctuating transport costs. By 1980, the value of beef exported as carcases represented over 80 per cent of the total cattle exports.

Winter wheat with tramlines to assist in spraying operations and reduce damage to the crop.

DARD

By the mid-1970s, the area of cereals had levelled off to around sixty thousand hectares, most of the grain being fed to home-based livestock. In GB, in 1978, there was a so-called 'Ten Tonne Club' consisting of cereal growers who had achieved a harvest of ten tonnes of grain per hectare. Ulster farmers were striving to match this performance, although the use of good seed, accompanied by judicious application of fertiliser, was required if such a yield were to be achieved. By the end of the 1970s, great interest was being shown in 'tramlining' of cereal fields. This referred simply to leaving unplanted tramlines or wheel ways through the crop when it was being drilled. They were laid down to suit the track and tyre width of the tractor wheels so that they could be used for subsequent field operations. The tramlines allowed more uniform application of fertilisers and sprays, with less damage to the crop by the wheels.

Potato harvesting by hand

DARD

Mechanical handling of potatoes resulted in less labour-intensive production.

DARD

From 1971, the area of potatoes grown in Northern Ireland each year had been declining almost continuously. Following the shortage in 1975, however, potato prices increased dramatically, and this stimulated a 30 per cent increase in the area of potatoes grown – from 10,900 to 18,800 hectares in 1977; but this fell again to 12,500 hectares by 1981. Pentland Dell was by far the most popular variety of potato in the late 1970s. With a few exceptions, vegetable production met the main needs of the home market.

Mechanisation was also on the march in the 1970s. Tractor numbers increased further and many of the operations carried out twenty years earlier by a man, a horse and a cart were now taken over by the tractor. Liquid manure spreaders were needed for the disposal of slurry from livestock systems. By 1981, annual investment in tractors, vehicles and equipment had reached £33 million.

The summer of 1975 brought new problems for farmers in Northern Ireland. May and June were the driest months for almost a century, and many farms were severely hit by drought. Water deficiency in the soil stunted the growth of grass, and silage cuts were lower. Hay crops were also lighter and, in County Fermanagh, the harvest was down by one quarter on previous years. The hot dry spell also took its toll on vegetables and grain crops. There was a drop in the potato crop, resulting in potatoes retailing at 14p per pound, 5p more than normal. Congregations in churches across Northern Ireland prayed for rain! Fears of a winter fodder shortage, however, proved largely unfounded. Because of the better weather, farmers were able to cut more, if lighter, hay, and more was being offered for sale. During July 1975, hay was making as much as £1.80 a bale ex-field, and few, if any, were quibbling about the price. Western Europe as a whole, including Northern Ireland, was in the grip of another unprecedented prolonged drought during the late summer of 1976. Problems again arose for livestock farmers because of the very poor growth of grass. Supplementary feeding had to be given much earlier than in a normal year. Some cows had to be culled to save fodder.

Changes in the Pigs Sector

The pig population continued the decline that had begun in the previous decade. A recorded figure of 698,000 pigs in 1976 (almost the same as in 1980) revealed a fall of 33 per cent on the figure for 1972, when pigs numbered over one million. Like other sectors, pig producers faced sharp increases in the price of feeding stuffs in the early 1970s. This had the effect of reducing profit margins and of curtailing further expansion. The Bacon Stabilisation Scheme helped pig producers until about 1971, but reduced prices after that resulted in a fall to 110,000 breeding sows by the end of 1973. The pig industry had reached its lowest ebb at the beginning of 1977. Census results in 1981 recorded the level of the pig-breeding herd as 65,000 sows and gilts – the lowest for well over twenty years.

The committee set up in 1968 under Professor John Ashton's chairmanship to examine the marketing of pigs in Northern Ireland, reported in 1970. The recommendations were far-reaching. The report recommended that the PMB should concentrate its operations on the production and marketing of pigs rather than on the processing and sale of pig products. It stated that the board monopoly and the quota system of supply to curers should cease and be replaced by a direct contract system between producer and processor. It recommended also that the board should liquidate its partnership interests in the processing industry. Should the board wish to retain some interest in the processing industry, the report said that this should be limited to not more than one wholly-owned factory. The report also implied that too much emphasis had been placed on bacon production (to be shipped as Wiltshire side bacon to GB) to the exclusion of other pig products such as pork. The recommendations provoked considerable debate, the UFU regarding the report as an attack on orderly marketing. Many of the recommendations were implemented despite strong resistance. The PMB made valuable efforts to bring about greater diversification of pig processing so that Northern Ireland producers had other products to sell as well as bacon.

An Evolving Poultry Industry

The egg industry was thriving at the beginning of the 1970s. Annual egg production stood at over 150 million dozen in 1970, a total that was more than 40 per cent up on figures for 1965. The BEMB, which had the general duty of improving the marketing of eggs throughout the UK, released weekly reports, detailing information pertinent to the industry. It also produced a quarterly review which gave details about supplies and prices of eggs not only in the UK but also in other EEC countries. After 1973, EEC regulations on marketing standards for eggs applied throughout the UK. All business premises selling eggs direct had to be registered as packing stations.

Egg producers faced a difficult time in 1975. Costs went up and egg prices failed to maintain the 1974 levels. The egg-laying flock on all holdings stood at almost 9 million birds in 1971, but this had fallen to 4.9 million by 1981. Over this period, the estimated average yield of eggs per hen increased from 218 to 249. During 1981, the output of hen eggs for human consumption was

Gosford poultry testing station
where commercially available
laying birds were assessed

DARD

95.4 million dozen, valued at about £34 million, over half of which were
shipped for sale in GB.

The marketing of eggs underwent a complete transformation at this time.
The BEMB since its inception in 1950 had provided an assured market for table
eggs at the ruling guaranteed price. The board was replaced in 1971 with a
non-trading Eggs Authority whose main responsibility was to improve egg
marketing. The result was that egg producers were, for the first time in thirty
years, selling their product on a free market. Egg prices showed little improvement, and many egg producers went out of business in the years that followed.
Pressures on the industry forced those producers who remained to join commercially integrated groups. These groups were very successful in achieving
advantages of scale in purchasing feeding stuffs and other materials, in
promoting efficient production techniques and in marketing effectively.

In Gosford, in County Armagh, the Department of Agriculture still
operated a testing station which demonstrated the relative qualities of commercially available laying poultry. New hybrid strains of laying birds were arriving
regularly during the 1970s, mainly from North America and GB. In the
Gosford random sample test of 1970, all commercial layers were hybrid strains.
That particular test had the highest hen-housed average of total entries ever
recorded to date – 242 eggs per bird. The test was won by an entry of 'Shaver
Starcross 288' hens with a hen-housed average of 268 eggs per bird in a
50-week laying period. This illustrated how productivity was increasing by
leaps and bounds through the introduction of improved hybrid strains. At the
same time, through higher efficiency, the number of birds in the broiler flock
continued to increase despite the end-price showing little change. In 1981,
30.5 million broilers were produced at a value of almost £32 million.
The principal export outlet for poultry meat from Northern Ireland was GB.

Continued Livestock Improvement

The AI service had become a major instrument in livestock improvement. The Department of Agriculture decided to reduce the size of the AI bull stud and to place greater reliance on semen from proven animals standing at various centres throughout GB. Semen from two categories was now available to farmers. They could opt for 'bull of the day' (BOD) semen at normal charges, or for semen from élite animals at higher rates. The demand for exotic Continental breeds which began in the late 1960s continued during the early 1970s. Breeds being used in Northern Ireland about the mid-1970s included the Blond d'Aquitaine, Charolais, Chianina, Limousin, Maine Anjou and Simmental. The Department of Agriculture and farmers were evaluating these breeds for ease of calving, growth rate and carcase quality. The view in the 1970s was that all of the breeds seemed capable of improving carcase quality of Northern Ireland beef cattle. Most also improved growth rates, but only the Charolais and Simmental breeds effectively colour-marked their progeny. None of the breeds possessed all of the desired characteristics, and there was often as great a variation within breeds as between breeds. They would, however, continue to make a valuable contribution to beef production in the decades that followed.

The sheep sector was also thriving. Breeders were continually encouraged through subsidies and grants to use high-quality rams in their flocks. The Ram Subsidy Scheme, introduced in the 1950s, was still operating, enabling breeders to purchase breeds such as Blackface, Border Leicester, Dorset Horn, Hampshire Down and Suffolk rams from leading flocks in GB. The Sheep Improvement Scheme also encouraged producers to maintain quality breeds, and grants were available to aid the purchase of Border Leicester rams to be placed in Blackface flocks.

A Pig Recording Scheme, which had developed out of the Pig Production Development Committee of the 1960s, obliged élite breeders to have their animals tested at the central testing station. Multiplying breeders were encouraged to produce larger numbers of cross-bred female stock for sale, and the scheme was also designed to ensure that commercial pig farmers had access to superior breeding stock.

Animal Health

Animal health continued to be a focus for attention, and Northern Ireland had a largely successful record in the 1970s. The old adage 'prevention is better than cure' applied very forcibly to the livestock industry. Major animal disease outbreaks were rare and this resulted, in large part, from the discipline and vigilance of the veterinary division of the Department of Agriculture. There had been a small outbreak of clinical Newcastle disease amongst exotic birds in May 1970, the first since 1949. The last major outbreak of the disease, also known as fowl pest, was in November 1973. The disease was finally eradicated at the end of December of that year, after a total of thirty-six outbreaks. A policy of slaughtering, cleansing, disinfection and vaccination of susceptible stock had taken place in the interim period.

Efforts continued in the 1970s to eradicate brucellosis. The Minister of Agriculture in 1970, the Right Honourable Phelim O'Neill, stated that there were now 'only about 150 herds with active infection. This is very encouraging but the final stages will require vigilance and co-operation of herd-owners.' In December 1971, the level of brucellosis had been reduced to a level at which Northern Ireland could be declared a 'brucellosis free area'. Subsequently, however, the level of infection crept up again, to 0.7 per cent in 1974. The EEC directive that governed trade in live animals required member states with a level of infection not below 0.2 per cent for two years to carry out the pre-movement testing of all animals over twelve months of age. This was a powerful incentive to eradicate brucellosis from Northern Ireland. Little did the Minister know that persistent efforts would be required to control brucellosis for many years to come.

Farmers and government bodies were particularly worried about rabies. To prevent infection reaching Northern Ireland, strict controls were placed on the importation of dogs, cats and certain exotic animals. These animals had to be quarantined for six months before being permitted entry. There was also debate in the early 1970s about the means of controlling foxes, since there had been a scheme of bounty payments for their destruction. With the changing methods of managing poultry, attacks by foxes were less likely, and reports suggested that more sheep were being killed by dogs than by foxes.

EEC accession highlighted differences in the animal health regulations between the UK and other EEC countries. One of the main differences was that the Community employed slaughter and vaccination policies against FMD, whereas the UK relied on a slaughter policy alone. This difference in approach was to raise its head again in the very serious UK outbreak of FMD in 2001.

Developments in Agricultural Education

With EEC entry imminent, there had been much talk about agricultural education. It was a widely held view that future prosperity depended largely on the calibre of those engaged in the agricultural industry. Farmers, farm

Students engaged in laboratory work at Enniskillen Agricultural College

DARD

workers, advisers, research workers and many others had a part to play. Some countries, like the Netherlands and Denmark, had a long tradition of agricultural education and training, which had helped to improve their competitive positions in Europe because of the highly efficient farming industries in both countries. Northern Ireland did not have as good a record. Just one in every three farmers had taken advantage of the facilities available for agricultural education and training, leaving a handicap which it was felt would take some time to overcome.

Reflecting the changes going on within the agricultural and food industries, Loughry became known as Loughry College of Agriculture and Food Technology from 1972. Foods that were prepared, packaged and branded ready for the oven were becoming popular with housewives. With an increasing proportion of food being consumed in processed convenience form, either ready to cook or ready to eat, Loughry College also took steps to provide personnel trained in food processing. The first food-processing courses in poultry technology had been introduced in the 1950s. A decade later, diploma courses in milk technology were started. From 1973, new technology courses at ordinary and higher national diploma level were introduced in commodities other than milk, such as meat and horticultural produce.

As the 1970s advanced, horticulture was making strides too. Expansion meant that worthwhile and rewarding careers concerned with the culture and care of plants were starting to appear. Opportunities in commercial horticulture included the production of fruit, vegetables, trees and shrubs. Amenity horticulture also offered jobs dealing with the construction and maintenance of parks, open spaces and recreational areas. Greenmount Agricultural and Horticultural College demonstrated in 1973 that it was moving with the times by providing a new course on amenity horticulture.

Another timely development in the agricultural education field was the introduction in 1972 of courses in food science in the Faculty of Agriculture and Food Science at Queen's University Belfast. With an ever-expanding world population and a growing shortage of food, agriculturists have made an enormous contribution to the resolution of food-production and processing problems. The faculty celebrated its golden jubilee in 1974. Since the first enrolments in 1924, approximately six hundred graduates had finished with degrees. Many stayed to play a full part in improving agriculture within Northern Ireland. However, many also went abroad, contributing to the advancement of agriculture in the developing world. The setting up of the Faculty of Agriculture in 1924 had indeed been a far-sighted and important development which had had a huge impact in agricultural education and research since its establishment.

Horticulture

In the 1970s, the horticultural industry began to thrive and to attract more substantial investment. The value of output in this sector had trebled in the previous ten years, and was running in excess of £12 million a year. Hybrid strains of vegetables, together with the use of herbicides, were providing higher yields. Government grants became available – the Horticulture Capital Grant

Scheme came into operation in January 1974 and remained open for applications until the end of 1980. Its function was to encourage growers to modernise horticultural enterprises, including land improvement and the construction and improvement of certain buildings. By the end of March 1978, almost 600 grant applications had been received. Some 474 were approved and brought benefits to applicants to the tune of £1.6 million. One major result of this grant-aid was the replanting of many long-established orchards, mostly in Armagh, during the 1970s. Northern Ireland was, and still is today, an important centre for the production of Bramley cooking apples.

Mushrooms had also become an important crop. Production started on a small scale and gradually expanded until output reached over 12 million pounds by 1972. In the mid-1970s, mushrooms accounted for about a quarter of horticultural output, and the industry supplied about one tenth of the UK's requirements. About thirty large-scale growers supplied the bulk of the market. The marketing of mushrooms had become a sophisticated operation, with large container lorries delivering the crop very quickly to markets in Glasgow, Edinburgh, Manchester, Newcastle and Liverpool. The combined output of field and horticultural crops accounted for just over 11 per cent of the total value of agricultural output in Northern Ireland between 1975 and 1976.

Great emphasis was placed on agricultural and horticultural co-operation in the 1970s. The Ulster Agricultural Organisation Society (UAOS), the central representative body for co-operatives in Northern Ireland, was responsible for the servicing of upwards of fifty affiliated trading societies, boasting a combined membership of almost 25,000 farmer members. By the late 1970s, these

Bramley apple production in County Armagh

DARD

societies were said to represent a total capital of around £16 million and an annual turnover in excess of £58 million. The Central Council for Agricultural and Horticultural Co-operation, a UK-wide organisation, had a regional officer based in Portadown. The council used the UAOS as its Northern Ireland agent in certain matters. From the mid-1970s, the government gave assistance for the development of marketing through a grant administered by the central council.

Afforestation

Up until the 1970s, forestry never really featured in agricultural returns or census reports. In comparison with other European countries, Northern Ireland was particularly lacking in trees, with only 4 per cent of its surface area covered. It was acknowledged that forestry was a sector in need of encouragement, as it could provide valuable raw material for future industrial development. Ireland was one of the least afforested countries in Europe, so there was much work to be done, not only in terms of planting but also in obtaining additional land. Some 50,000 hectares of forest had already been established and the planting rate was about 1,000 hectares a year. The aim was to reach the combined target of 120,000 hectares set by private and state forestry for the year 2000. The market for home-grown timber had two outlets – sawmills and a chipboard factory at Coleraine.

In the late 1970s, over 80 per cent of new forests were planted on low-agricultural-value land west of the Bann and in parts of north Antrim. Thriving forests were established with the help of sophisticated ground-preparation machinery. The introduction of the Sitka spruce from the west coast of America produced more trees quickly under extremely poor conditions. The establishment of lodgepole pine and Japanese larch also contributed to the development of forestry in Northern Ireland – each in its own way. In the interests of conservation, great care was taken to ensure that forests being planted blended in with the landscape of the area. Pomeroy Forestry School provided courses for over five hundred full-time forest workers and supervisors.

Forestry development was also an opportunity to bring maximum benefits to the community in the form of wildlife conservation, recreational potential and educational opportunities. There were wildlife areas where shooting was not allowed, and grouse moors were set aside to enable the red grouse to survive. At Seskinore Forest in County Tyrone, thanks to stalwarts such as Jim Whiteside, an extensive farm operated, raising some twenty thousand game and wildfowl for release each year into forests across the country. There was a growing demand in the 1970s for certain specialised activities, such as camping, pony-trekking, and caravanning. From 1967 and during the 1970s, a number of 'Tree Weeks' were organised, during which schools, villages and towns co-operated in the ceremonial planting of trees. These events were successful in stimulating greater interest in tree planting and also in creating and fostering an awareness of the importance of trees in the countryside.

Conservation ponds as a step towards improving the rural environment

DARD

Farming and the Environment

The 1970s began with what was described as 'European Conservation Year'. This was organised under the guidance of the Council of Europe and was aimed at getting people from all walks of life to take a pride in improving their surroundings. The farming community obviously had a responsibility to play its part in improving the natural beauty of the landscape. There was often a conflict between farming as efficiently as possible and conservation. For example, the mechanical aids being used in farming, such as combine harvesters, necessitated larger fields and fewer hedges. Nevertheless, farmers were encouraged to retain rough grazing or even clumps of trees in corners of fields. These measures went some way towards providing cover for birds and smaller wild animals. It is worth recording, however, that by the end of the decade, the characteristic call of the corncrake had virtually disappeared from the Northern Ireland countryside. There is little doubt that the practice of cutting grass for silage in early summer was a contributory factor in its decline.

Pollution was another area that continued to cause problems in the 1970s. The increasing numbers of stock and their concentration into larger units resulted in slurry and effluent-disposal problems. Conservation of grass as silage was still increasing, resulting in more silage effluent and greater potential for serious pollution. Self-feeding and the use of cubicles for cattle also meant that the volume of traditional farmyard manure, based on straw bedding, was reduced considerably. Pig houses with slatted dunging passages above collection tanks were increasing in number each year. All of this pointed towards an even greater volume of liquid manure and effluent that had the potential to escape into unwanted areas.

The result was that rivers and watercourses were still being contaminated with farm wastes including silage effluent and residues of pesticides, fungicides and herbicides, although factories were often responsible too. Fish deaths occurred over long stretches of rivers as a direct result of escaping effluent. Farmers spent considerable sums building slurry tanks and manure pits to hold effluent produced by farm animals or coming from silos. On average, about ninety farmers per year in Northern Ireland were being prosecuted for causing pollution. Although fines of a few hundred pounds were often imposed, there were also additional fisheries-reinstatement costs which farmers invariably had to pay.

When speaking at the Scott Robertson memorial lecture in November 1972, James A. Young, then Permanent Secretary of the Ministry of Agriculture, said with great foresight:

Farmers must get used to the idea of increasing numbers of townsfolk coming out to enjoy the pleasures of the countryside during their leisure time. Farmers may even be irritated from time to time by townsfolk who are more zealous than tactful in their attitude to matters rural. For example, in future more people rather than less may voice their complaints about the appearance of farm buildings or about the welfare of animals kept intensively or about the removal of hedges.

Farming Organisations

As the 1970s came to an end, the many organisations that supported Northern Ireland agriculture were working well. The UFU, despite the formation of the Northern Ireland Agricultural Producers' Association (NIAPA) in 1974, was still acknowledged as the main voice of farmers in Northern Ireland. The farmers' union had been formed as far back as 1917 and had been involved in almost every major farming policy in the intervening years. It boasted a membership of seventeen thousand in 1980 and was closely allied to its 'sister' bodies in England, Scotland and Wales. They met continually to negotiate in consort with MAFF at UK level. The UFU had some fifty field staff with central offices in most of the main market towns in Northern Ireland. The UFU offered advice on a day-to-day basis on a plethora of problems ranging from drainage matters to insurance issues. At its headquarters in Belfast, a team of specialists represented farmers on a wide range of committees.

The Young Farmers' Clubs of Ulster was formed in 1929. Under the motto 'Better Farmers, Better Countrymen, Better Citizens', the movement continued to promote cultural, educational, recreational and social activities. Its comprehensive programmes of competitions over many years made a considerable contribution to the social and educational development of young people throughout Northern Ireland.

The RUAS, founded in 1826, organised its annual agricultural show each year. The show was one of the most important events in the farming calendar. It normally took place over four days each May, but it had an enhanced profile in the last half of the 1970s. In 1976, the Granville Nugent Hall, a new exhibition building of around twelve thousand square feet was completed, to meet the demand for extra exhibition space. This gave the society five major halls at Balmoral for staging exhibitions and demonstrations. Exhibitions during 1978 were of record proportions, consolidating Balmoral as the number-one showplace in Northern Ireland at that time. In addition to its encouragement of agriculture and industry, the RUAS was also charged under its constitution with the promotion of the arts. In 1972, the Harberton Theatre was opened, soon becoming a well-established centre for light opera and pantomime. Musical festivals were also the order of the day in the E.T. Green Hall of the Members' Rooms.

The Northern Ireland Agricultural Trust

The position of farmers in Northern Ireland received a boost with the formation of the Northern Ireland Agricultural Trust in March 1967. The trust had a small, highly experienced staff team with much specialist knowledge of agricultural production, food processing and marketing. Funding was provided initially from the Remoteness Grant. The trust's function was to assist in the development of projects thought likely to improve farming income. In particular,

The Northern Ireland Agricultural Trust encouraged the development of experimental projects, such as flax growing, which might have potential on Northern Ireland farms.

DARD

its aim was to foster developments which lay outside the scope of the Ministry and which were not yet at the stage of private commercial operation. The trust was empowered to conduct investigations and to invest in a wide range of projects. It organised and financed many experimental projects and developments in agricultural production and marketing for which commercial finance would have been difficult to find. Such projects included flax processing, grass drying, potato processing, cattle shipping, gravel tunnel drainage and many more.

The trust encouraged and co-ordinated the participation of Northern Ireland interests at national and international food and trade fairs, exhibitions and promotions. In the 1960s and through to the late 1970s, the trust played a crucial role in promoting Northern Ireland produce in both domestic and international contexts, particularly in Europe. It acted as a shop window for Northern Ireland produce on visits to Dublin and Edinburgh, and when it attended major food fairs in Paris or in Cologne – the world's largest.

Towards the end of 1979, however, the then chairman, Dawson Moreland, had the sad task of revealing that the Trust was to go, the government having decided to cut off the grant under a quango-cutting exercise. Moreland deemed the decision 'a bad one, based on an error of judgement'. It even occasioned North Down Unionist MP, James Kilfeddar, to accuse the government of 'strangulation of the Trust in pursuit of ruthless public expenditure cuts without regard to the special conditions existing in Northern Ireland'. The Northern Ireland officer of the Irish Congress of Trade Unions, Terry Carlin, called on the government to reprieve the trust. He said that his organisation was 'deeply dismayed' at the decision. When the end came, through a Commons Order late in February 1981, the curtain came down on what had been an extremely exciting era. During its short existence, the trust contributed much to the Northern Ireland agricultural industry.

Ulster Farm in the 1970s

Ulster Farm continued to operate its rendering business at the Glenavy site throughout the 1970s. Additional equipment was purchased, enabling the factory to process blood into blood meal. During late 1971, new raw-material intake equipment was installed. A major review of the whole production process was carried out and, following extensive research involving visits to other plants in Denmark and Sweden, new 'state of the art' equipment was purchased. It had three main features – pressure sterilisation; production of high-protein meal with low fat content; and production of high-quality tallow. This system worked on a batch cooking/sterilisation principle, with liquids and solids being separated by centrifuge and decanter. The resultant meal was dried in small continuous disc driers. After initial commissioning difficulties, this plant gave good service until 1984 when the industry throughout Europe was again adopting different technology.

The transport fleet was upgraded in the 1970s, including enclosed skips and articulated vehicles. Capacity at Glenavy increased from about 500 tons to 1,300 tons per week, with the equipment running continuously. This meant, of course, that there was almost a three-fold increase in the product for sale. The meat and bone meal was of good quality because of its high protein content, and was mainly sold locally to the milling trade.

One troublesome incident occurred on 27 May 1973. A terrorist bomb was placed between two cookers at the Glenavy plant; fortunately it did not cause the damage it might have done. However, a lot of hard work was required to resume production again with the minimum of delay.

Ulster Farm board members pictured with company secretary, William McLornan and pork consortium representative, Wilson Robinson, in 1979. Left to right: Geoffrey Conn, William McLornan, George Ervine, Eddie Swain, William Wilson, William McCollum, Wilson Robinson, Crawford Wadsworth and Verdun Wright.

Douglas Higginson

Conclusion

The UK achieved full membership of the EEC in 1978, following the agreed five-year transition period, and the agricultural support system was fundamentally redirected. This was the most influential development of the 1970s. Previously, in the UK, deficiency payments, grants and subsidies had been employed to provide food for consumers at a relatively low price. The government made up the difference between the imported price and that negotiated at each annual review, hence the term 'deficiency payment'. By this means, the general public obtained its food at reasonable prices, while the farmers were able to make a reasonable living. This system worked satisfactorily at the time but was incompatible with that operating in the EEC. Under the CAP, farmers were guaranteed returns close to the prices paid by consumers. This was achieved through an 'intervention' system, where a standard intervention price (essentially a 'bottom' to the market) was fixed for each commodity. Farmers were also paid direct grants for specific purposes, such as assistance given to farmers in the LFAs. The trouble was that the prices paid to farmers were higher than those outside the EEC, and tended to encourage over-production. This was to lead to problems in the next decade.

Developments in housing, management, feeding and breeding of most farm stock in the 1970s meant that feed-conversion rates improved, yields increased, production became concentrated into even larger units, and labour was used more efficiently. A larger proportion of farm animals was being housed inside for a greater part of the year. Developments in crop production, such as the use of 'tramlining' and the further use of chemicals for disease control, also contributed to higher crop yields. In successive years, 1975 and 1976, farming was adversely affected by very dry weather, with lower crop yields and higher prices. At the end of a decade that had seen the agricultural industry almost bankrupted by the 1974 slump, things were looking a little better for the industry and those who worked in it.

1980s

Introduction

One of the major decisions affecting Northern Ireland farming in the 1980s was the introduction of milk quotas. Although opposed initially by farmers throughout the EEC, quotas were later to be a very important asset in the balance sheet of every dairy farmer. The management and pattern of milk production were altered by the imposition of quotas, as farmers tried to reduce costs and tailor production to their entitlements. During the 1980s, the move towards a smaller number of larger dairy holdings continued. The traditional mixed-farming pattern was being replaced by still more specialisation. The pig industry continued to decline, and egg production remained in the doldrums. However, the poultry-meat industry grew because of the significant marketing and product-development efforts put in by firms such as Moy Park and O'Kanes. During the same period, agricultural co-operatives came under great financial pressure because of the lack of capacity utilisation and the unit cost of processing.

Animal disease, as in previous decades, was never far from the headlines. Brucellosis and tuberculosis continued to present difficulties in their control, and mastitis in dairy cattle remained a problem. But the most significant development in the late 1980s was the recognition of the disease known as BSE – Bovine Spongiform Encephalopathy. This had particular implications for Ulster Farm, leading to the introduction in 1989 of new regulations concerning specified risk material. BSE, together with FMD, was to assume major significance in the period ahead.

Nearly all of Northern Ireland's agricultural exports continued to be sent either to GB or to the Republic of Ireland. Seed potatoes were one exception, and a significant proportion of each year's crop was shipped to many Mediterranean countries, to Egypt and to the Canary Islands. Bacon, ham, and meat products were regular exports, as were vegetables, apples, mushrooms and processed products. In 1991, the total value of Northern Ireland's agricultural exports at farm-gate prices amounted to £548 million. Agriculture remained a major player as far as employment was concerned. Almost 40,000 farmers toiled on the land with the assistance of another 19,000 employees working alongside them.

Environmentalists were publicising their concerns more than ever before, and farming had to take account of their views. Farmers were also forced to give more recognition to the views and demands of their customers, who were taking greater interest in the quality and presentation of food, and whose tastes in food were changing. Food safety was also becoming an issue towards the end of the 1980s, prompted largely by the BSE crisis.

EEC Membership

By the 1980s, farming in Northern Ireland was dominated by membership of the EEC (or EC – European Community as it was then known) and by the CAP. Leading farmers and civil servants were called upon to attend EC meetings in Brussels and Luxembourg almost weekly. Many farmers in Northern Ireland felt that farmers on the other side of the border had a better deal, because the Republic of Ireland was a member state in its own right within the EC, while Northern Ireland was only part of a member state. Northern Ireland's priorities were rarely the same as those of the rest of the UK.

Early in 1982, the Common Market was plunged into its worst crisis in sixteen years. Poor markets and falling prices for many commodities took their financial toll on the farming industry – when Britain's Community partners over-ruled the government's veto of record EC farm price increases. Agriculture Minister Peter Walker had demands for a majority vote imposed upon him when the other member states called for an 11 per cent farm price increase. He boycotted the voting procedure, believing that it breached the so-called 'Luxembourg compromise' of 1966 which had brought in the unanimous vote system.

In 1983, the new Northern Ireland Assembly was also showing its teeth as far as agriculture was concerned. Members took part in a three-hour debate at Stormont, during which Secretary of State James Prior and his team were strongly criticised for their 'failure to recognise the plight of Northern Ireland's farming industry'. Members appointed a five-strong delegation to confront Prime Minister Margaret Thatcher at Downing Street. The delegation included Dr Ian Paisley, John Taylor, Addie Morrow, Simpson Gibson and James Molyneaux.

By the middle of the 1980s, another problem was looming. A bad EC beef-price deal looked set to force farmers to sell cattle in the Republic of Ireland where they would get better returns. This created fears that meat plants in Northern Ireland would have to close since material was being sold to the Republic. The price deal took over seventeen hours to hammer out, but, at the end of the negotiations, Northern Ireland farmers felt that they had made little progress. Mr Brendan McGahan, chief executive of the LMC, described the negotiations as 'disastrous for the Province's beef producers'.

Milk Quotas

It became clear in the 1980s that farmers' success in producing food was beyond the ability of countries to dispose of it economically, although people in many parts of the world still went hungry. Continuing over-production and disposal of surplus milk had been a serious problem in the late 1970s and early 1980s. The 'intervention' system of supporting farm prices had created 'mountains' of surplus food, especially dairy products and beef. The cost of disposing of the UK surplus milk as butter and skimmed-milk powder was running at some £3 billion per year in 1983. It was obvious that such an open-ended financial commitment would be difficult to sustain as additional milk surpluses built up. Drastic action was required. A system of quotas based on past production was the only way forward that was acceptable to the Council of Ministers. On 31 March 1984, a quota system was formally adopted in the milk sector. Milk production quotas were imposed on each member state, with a supplementary levy (referred to as a 'superlevy') at a penal rate applied to any milk produced above the quota. This meant that if Northern Ireland as a whole was below its quota, no levy would be imposed. However, if milk production exceeded the quota, a superlevy would be charged.

For 1984, dairy farmers in Northern Ireland were allocated a quota that was 9 per cent below the 1983 milk-production levels. Milk producers were anything but happy about their situation. They faced the prospect of having to reduce production as part of the EC plan to fix reduced national levels for milk. Dr George Chambers, managing director of the Northern Ireland MMB, said that the deal left the industry facing significant uncertainties. The situation was alleviated slightly by the decision that Northern Ireland's 9,000 dairy farmers would be permitted an extra 65,000 tonnes of milk – equivalent to about 63 million litres – on top of their quota.

The extra allocation, however, became a bone of contention when farmers' leaders accused the government of 'swindling' them out of it. Representatives of the MMB, the UFU, and NIAPA, accompanied by Northern Ireland's three Members of the European Parliament (MEPs) went to meet Agriculture Commissioner Paul Dalsager in Strasbourg. On their return, Gabriel O'Keefe, chairman of NIAPA, said that they had been assured that the government would be expected to explain in the fullest terms how the quotas were allocated. Agriculture Minister Michael Jopling produced figures for Brussels, purporting to show that Northern Ireland had received its full 65,000 tonnes of extra milk quota but Dr Chambers contended that the special aid for Northern Ireland had been distributed to England, Scotland and Wales.

In the light of milk quotas, all producers were forced to assess their position and reach decisions on what to do, since the imposition of superlevies had the potential to be very severe. Expansion of milk production was no longer an option, as it had been in the past, although maximum profit could be made only if farmers filled their quota. Farmers had to find answers to questions such as whether to reduce the size of their herds, to lower concentrate feeding, to try to reduce or increase milk yields per cow or to alter the fertiliser application to their fields. There was also considerable debate about the circumstances when it would be sensible to buy quota or lease additional land with quota attached. The only way forward seemed to be to concentrate on improving profit margins by reducing costs, since increasing output was likely to be penalised. Milk quotas certainly stimulated producers to think fundamentally about their milk-producing policies, often necessitating adjustments to their previous methods.

As it happened, circumstances resulted in few or no superlevy penalties being imposed in Northern Ireland in the first three years of milk quotas. However, because of good conditions for milk production in 1987, superlevy payments had to be paid by about one third of milk producers in 1988. About 2,700 producers in Northern Ireland paid approximately £2.7 million in superlevy in that year. One consequence of the introduction of quotas was that milk smuggling began to operate across the border into the Republic of Ireland. As Dr Chambers pointed out: 'Milk is now being smuggled across the border into the Republic where it is being sold at 20p a gallon cheaper than the board would pay, so that farmers can avoid paying a superlevy which is imposed for overproduction.' Following the introduction of milk quotas, the size of the Northern Ireland dairy herd fell by twenty thousand cows in the four years to 1988.

The weather has always been an important factor in farming – hill ewes in the snow

DARD

Farming Income and the Weather

At the start of the 1980s, interest rates had eased and the battle against inflation was gradually being won. After a poor year in 1980, farming income for 1981 recovered to almost £55 million, mainly because of higher producer returns. 'Farming income' refers to the income from Northern Ireland as a single national farm. Because of a rise in the value of milk sales, 1982 was also a more rewarding year, and 1984 was one of the best years for over a decade. But by 1985 and 1986, there had been a significant drop in farm incomes, although there was a considerable recovery towards the end of the 1980s.

Bad debts were causing concern in the mid-1980s, particularly among agricultural machinery dealers, who were being forced to adopt a 'cash only' policy when providing spare parts or service. Mr Holmes Haslett, chairman of the Northern Ireland Agricultural Machinery Dealers' Association, knew of small dealers who were owed more than £100,000 by farmers. Banks also feared that some farmers had overstretched their credit limits and were anxious about the amounts owed to hire purchase and leasing companies. The Department of Agriculture estimated that in 1989, Northern Ireland agriculture had an interest bill of £48 million, based on an average bank-interest rate of 17 per cent on £285 million owed by farmers to the four main local banks. The financial squeeze was clearly exacerbated by the high interest rates involved in borrowing money.

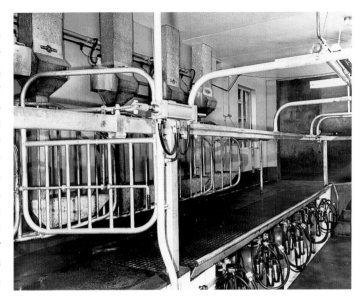

Heavy investment in farm buildings and equipment could result in excessive debt, particularly when commodity prices were low.

DARD

As always in farming, the weather can upset the best of plans. The years 1980 and 1981 brought poor weather, but worse was to follow later in the decade. The wettest summer that many could remember occurred in 1985, creating tremendous difficulties for farmers in Northern Ireland. With constant rain over the summer period, silage making was extremely difficult, and haymaking almost impossible. For the first time in thirty years, farmers had to import hay to feed their winter livestock. The potato crop was also badly affected, and the Assembly's Agriculture Committee Chairman, Dr Ian Paisley MEP, along with other members of the committee, visited a south Down farm to see at first hand the results of the bad weather for potato farmers. 'What I have seen would bring tears to your eyes,' said Dr Paisley. 'Many of these farmers will have to go out of business.' Potato production took some time to recover from the effects of that summer.

Social Democratic and Labour Party leader John Hume MEP told the European Parliament in Strasbourg that some farmers had been driven to suicide as a result of the financial ruin brought about by the disastrous weather. He and Dr Paisley called for special aid to help farmers who had lost up to £40 million because of the bad weather. Sinn Féin also criticised the government's handling of agriculture in Northern Ireland. Party spokesman Jim McAllister was scathing in his attack on Lord Charles Lyell, the Minister dealing with Northern Ireland agriculture. The latter took to the fields and the farms to find out for himself the real extent of the situation. The bad weather also brought worries for those involved in the mushroom industry, which, in 1985, was worth over £9 million. Oliver McCann, specialist adviser to the mushroom industry, warned that one thousand jobs could be lost. The lack of straw was the industry's major problem. It required some forty thousand tonnes of straw a year to keep the industry adequately supplied with compost.

The government did provide some aid for weather-beaten farmers in Northern Ireland in the form of a £5 million package which enabled livestock breeders to enjoy increased HLCAs, and provided compensation for those who sustained losses in hay and silage because of the bad weather. The UFU gave the aid package a guarded welcome. It felt that the money, which to a large extent went to farms in LFAs, should have been available to all beef farmers. The spring of 1986 turned out to be the wettest of the century.

Cropping and Horticulture

Farmers were continuing to devote more land to grass production. One technique that was frequently used was to apply nitrogen to produce 'early bite', particularly for dairy cows, since extra milk and better growth rates could be produced more economically from early grass. Silage production increased yet further – in 1986, some 6.5 million tonnes of silage were made in Northern Ireland, more than in any previous year. Some of this from the mid-1980s onwards was made using the 'big bale' system, which was growing in popularity. New and improved machines for the baling, wrapping and sealing of wilted grass came on the market. The Department of Agriculture ran working demonstrations of big-bale silage at Greenmount, in 1988. Although some farmers considered big baling to be more expensive than pit silage, the system had a place on many farms during the 1990s.

The area devoted to arable crops declined further in the 1980s. There were now only two major arable crops – barley, largely grown for animal feed, and potatoes, to meet both ware and seed markets. More farmers were changing to winter cereals, although in the mid-1980s over 80 per cent was still planted in the spring. Farmers were advised that they needed to plant cereals early to get the best possible yield, with the last fortnight in March being a realistic target. The Department estimated that 'yield drops by half a tonne per hectare for every week's delay in sowing after the first of April'. Seed-potato growing declined much more than ware as the 1980s progressed, falling in value from £8 million in 1983 to £5 million in 1987, whereas the ware crop was valued at £15 million in each of these years.

In 1986, government research revealed that leatherjackets, grubs of the crane fly, were again causing damage to grass on Northern Ireland farms, estimated at £15 million each year. The culprit was well known in cereals, particularly spring barley, but was also ubiquitous in grassland. The Department of Agriculture had conducted surveys of leatherjackets for over twenty years and had found that there were over half a million leatherjackets per hectare, on average. A commercially available sampling technique became available in 1986, allowing the numbers in grassland to be estimated from November onwards, and insecticide applied.

The production of mushrooms continued developing during the 1980s. As a result of high oil prices and poor returns in the mid-1970s, mushroom output had declined. Its subsequent growth was one of the success stories of the agricultural industry, contributing significantly to the local economy. Its success was attributable to efficient production systems, a quality product and good marketing. There were lessons here for other enterprises! A survey carried out by the Department of Agriculture found that there were about 130 mushroom growers in 1982–83, 40 of whom were specialist growers, with no other enterprises. The average size of unit was getting bigger, reaching nearly ten thousand square feet by 1983. The survey showed that 9 per cent of mush-room growers produced 43 per cent of the total output. New production tech-niques such as pasteurised compost and bag growing were introduced, which required much lower capital than conventional tray production. The most important markets in GB during the 1980s were Glasgow, Liverpool and Manchester.

As the 1980s drew to a close, some consumers were querying the current methods being used for growing potatoes and vegetables. One alternative that was considered and implemented by a few farmers was organic farming, where no artificial fertilisers or chemicals are used in producing the food. Demand for organically grown food was increasing rapidly in GB. Because of lower yields and higher labour costs, this method of farming was more expensive than traditional intensive methods. The big problem was to identify and develop market outlets for the organically grown produce. Consumers had, of course, to be prepared to pay the higher prices of organically grown products if this method of production were to succeed. For this reason, expansion of organic food production is likely to be slow.

Animal Enterprises

Organically grown field vegetables

Greenmount College of Agriculture and Horticulture

Cattle and milk production were still the most important livestock enterprises. From the mid-1970s, the total cattle population had stood at between 1.4 and

1.5 million through most of the 1980s. In June 1981 and 1991, the numbers in the dairy herd were almost identical, at about 270,000 cows, although the number rose slightly above this in the mid-1980s. The MMB purchased virtually all available milk produced on over eight thousand farms, each of which had an average herd size of some thirty-six cows. Of the 1,418 million litres of milk sold, around 15 per cent was destined for the liquid market, while 54 per cent went for butter manufacture and around 20 per cent for cheese production. It was inevitable that this changing market situation would have an impact on the price of milk to the producer.

Although there had been considerable progress in previous decades, there were yet further developments in the management of dairy herds during the 1980s. Cubicle designs were altered to provide more comfort for the cows. Milking parlours moved on from low-jar systems to big-bore direct to line systems, and the herringbone type was still the most popular. Electronically controlled milk recording and feeding were introduced, allowing more precise management of dairy herds. Forage conservation methods also changed, with new additives and grassland management techniques producing higher quality fodder.

The beef cow herd increased from 205,000 cows in 1981 to around 250,000 in 1991. The demand for cattle for fattening had to be met by importing thousands of store animals from the Republic of Ireland. Sheep numbers climbed each year throughout the 1980s from about one million at the start of the decade to 2.5 million by 1991. The rapid expansion and increased profitability of the sheep sector was undoubtedly the result of the sheep meat regime introduced by the EC in 1980. It provided security and price support which contributed to the growth of the sheep flock, which growth was, no doubt, also helped by the imposition of milk quotas and the recurring problems of the beef industry.

The high-price regime adopted under the CAP had taken its toll on Northern Ireland's intensive livestock sector. Because of a fall in demand for eggs, there was a decline in the laying flock from 5.7 million in 1981 to 3.7 million by 1991. A salmonella scare, which received more than its fair share of media attention, eventually leading to the resignation of Secretary of State for Health Edwina Currie, in December 1988, probably contributed to the reduced demand for eggs. By the middle of the decade, some 5,000 farmers in Northern Ireland kept laying hens but 85 per cent of the birds were on fewer than 200 farms. In 1991, almost 63 million dozen eggs were produced, worth £29 million to producers. The average producer price per dozen was 40p in 1991. Almost half of the eggs produced were exported to GB. To a large extent, the drop in the laying flock was offset by a rise in the number of table poultry, mainly broilers, which exceeded five million at each annual census in the late 1980s. The majority of broilers and turkeys were reared on fewer than 150 farms. In 1991, output of poultry meat stood at some 82,000 tonnes and brought in £62 million, almost half of this amount being sold in GB. Pig numbers in the mid-1980s stood at around six hundred thousand, just over half the number recorded in the mid-1960s.

Diversification

With the opportunities for expanding traditional enterprises limited by surplus production and lower prices, farmers were looking at other possibilities to maintain or increase their income. Many suggestions were made, such as producing edible snails, farming goats and sheep for milk, producing venison, growing organic carrots, making baskets and becoming involved in tourism – some of these ideas being taken more seriously than others! But these often required different skills that had to be learned. In addition, the demand for some of these products was likely to be limited. Success sometimes also depended upon having access to a nearby urban population as a potential market. From January 1988, grant-aid for tourism and crafts was available under the Farm Diversification Grant Scheme. The new scheme provided assistance for farmers wanting to begin a range of alternative enterprises. Farmers had to put forward proposals as part of a potentially profitable plan, and grant-aid of 25 per cent was payable. Items eligible for grant-aid included the processing of farm produce, craft manufacture, farm shops, and catering and recreation facilities.

One viable crop, which was considered a reliable alternative to cereals for the Northern Ireland climate, was oilseed rape. It was introduced as a break crop in the early 1980s, and between 1982 and 1985 the area grown trebled to about five hundred hectares. Fields of the crop with its bright yellow bloom were very striking across the rural landscape. The crop was grown for the black cabbage-like seeds which, when crushed, produce oil for use in industry. Much of the initial growing of this crop was in the Limavady area. Rape seemed suitable to the soil and climatic conditions, and was capable of being combine

A crop of oilseed rape, often grown as an alternative to cereals when the economics were favourable

Esler Crawford Photography

Deer farming as an alternative farming activity

Michael Drake

A farm shop selling its own produce directly to consumers

DARD

harvested, although harvesting difficulties were sometimes experienced. It was profitable at the prices prevailing in the mid-1980s. In 1987, however, reduction in the support price reduced the attractiveness of rape, although it was still considered useful as part of a cropping rotation.

A survey was carried out by the Department of Agriculture in 1989 on how Northern Ireland's farmers had diversified in their use of land. Among the most frequent alternative activities were agricultural contracting, horse breeding, goats, farmhouse accommodation and the letting of cottages for self-catering. Although not in great numbers, unusual livestock or crops included deer, fish, rabbits and Christmas trees. Activities related to tourism or recreation included campsites or caravan sites, farm walks, nature trails, pony trekking, shooting and fishing. Some farmers had also diversified into direct sales of products such as eggs, fruit, meat and vegetables, some through establishing their own farm shops and selling direct to consumers. The Department advised in 1990 that 'quality vegetable production, pick-your-own fruit and propagation of shrubs, hedging plants or young trees are all worth considering, provided the expertise and market outlets are readily available'.

Agricultural Contracting

There was a steady growth in agricultural contracting throughout the 1970s, and this accelerated during the 1980s. Contracting had been accepted for operations such as lime spreading, hedge cutting and combine harvesting where very specialised, expensive machinery was required. The high capital outlay involved in purchasing such machines was not justified on many farms, making it sensible to hire contractors for some of the major farm operations. Contract charges were often lower than in GB, partly because many of the contractors were also farmers and were simply trying to utilise the spare capacity of expensive machines by working for their neighbours. The use of contractors represented a big change in attitude by farmers, many of whom were very independently minded. Economic pressures largely dictated the change.

A survey carried out by the Department of Agriculture in 1987 showed that about 13,500 farmers in Northern Ireland employed contractors for at least one operation on their farms. The contracting activities carried out most frequently included harvesting, drainage and site work, baling, hedge cutting and the spreading of slurry and manure. However, in terms of expenditure, baling was at the head of the list, followed by combine harvesting, slurry and manure spreading and various cultivations and hedge cutting. In 1987, the service was provided by an estimated 560 agricultural contractors, many of whom operated contracting alongside a farm business.

Arrival of Computers

Computers with special application for farming emerged in the 1980s. The cost was coming down and the decision to purchase a computer was not as major as it once had been. Personal computers offered the possibility of eliminating tedious manual calculations, as well as giving help with analysing

Farm management using computers

DARD

records. Suitable software programs were becoming available which could record financial and physical data on farm businesses, dairy and pig herds being particularly suitable. The development of information and management programs allowed farmers to keep a close check on enterprise performance. Accounting packages were also used to keep track of both expenditure and income, including monthly Value Added Tax (VAT) returns. Young farmers returning to farms from the colleges were more computer literate and were prepared to experiment with new technology. Finding a couple of hours each week to keep up with the bookwork was not always easy, and computerisation proved a means of lightening the burden. As farming entered the 1990s, it seemed that the need for records, information and budgeting was bound to increase, making the computer an inevitable tool on most sizeable farm businesses.

Control of Animal Disease

Although measures to control tuberculosis had been in operation for over thirty years with considerable success at the start, it was proving a very stubborn disease to get under control. Changes in farming practice, animal husbandry and stocking levels affected the spread of the disease, allowing it to become re-established again. Over three hundred herds were under restriction in 1982. The lowest incidence of tuberculosis since 1972 occurred in 1986, but by 1988 there had been an increase in certain areas. At the same time, the eradication programme for brucellosis had been very effective with Northern Ireland virtually disease free – a considerable advantage when exporting cattle. Edwin Conn, chief veterinary officer at the Department of Agriculture for over twenty years, played a key role in the eradication of bovine tuberculosis and brucellosis, and his efforts will be remembered. In 1982, for example, there were only 22 herds affected by brucellosis out of a total of some 45,000. After his retirement, Conn was made a Fellow of the Royal College of Veterinary Surgeons, the first such award in Northern Ireland since the 1960s. He also became a technical consultant to Newry-based Norbrook Laboratories after he finished working for the Department.

Swine fever, which had last appeared in Northern Ireland in 1958, was another serious disease that had to be kept at bay. In 1980–81, there were outbreaks of FMD in France, Jersey and the Isle of Wight, putting the livestock industry under red alert. Every effort was made to ensure that this devastating disease did not reach Northern Ireland which had been free from the disease since 1941.

Northern Ireland had a very low incidence of sheep scab for many years, but there was always a danger that it would gain a foothold. During October and November each year, all sheep were dipped for the prevention of sheep scab. The disease could have serious financial consequences through losses in wool, condition of the sheep, and even death in young lambs. Unfortunately, during the year ending March 1987, there was a sudden and alarming increase in sheep scab, with forty outbreaks. The Department introduced a second compulsory dipping period in an effort to control the disease.

The MMB and the Department made a concentrated effort to stamp out the insidious disease, mastitis, which was costing the dairy industry an estimated £6 million a year. In the previous fifteen years the average cell-count had been brought down from just over half a million per millilitre of milk in the early 1970s to a very low level of one third of a million per millilitre. But an estimated 30 per cent of producers still had a serious mastitis problem, with losses in milk yield averaging 360 litres per cow. The campaign emphasised the need for continuing vigilance.

Thousands of farmers benefited from a breakthrough in animal health-protection costs by a County Down pharmaceutical company. As a result of a £1 million investment, Newry-based Norbrook Laboratories claimed it could save farmers thousands of pounds in treatment costs for their animals' gastrointestinal problems. Edward Haughey claimed that intensive research had reduced the production costs of an anthelmintic that could be used to treat cattle and

Suckler cows and calves with clearly identifiable ear tags

Michael Drake

sheep. Edwin Conn endorsed Mr Haughey's comments: 'There is a need to protect animals reared here and at times it has proved to be costly. We have produced something that will save farmers money and no doubt they will take advantage of it.'

One advance which was to prove of considerable value in the 1990s and beyond was the animal-health computer project developed by the Department of Agriculture and installed in 1988–89. The Department had responsibility for the animal disease-eradication programmes, an essential element of which was the control of cattle movements in Northern Ireland. Animal test records and stock movements were maintained, and there was the facility to trace possible contacts with infected animals. Every beef and dairy animal in the country had an ear tag with a unique identifiable number. This cattle-tracking system had been operated manually since 1964, but the large amount of paperwork had become cumbersome. All of these records were fully computerised from 1988, conferring tremendous benefits on herd owners, veterinary practitioners and the Department, by providing accurate and up-to-date information on animal movements. The ability to trace cattle movements was to prove of considerable value in providing unique guarantees on the freedom of disease during the BSE crisis in the 1990s and the FMD outbreak of 2001. In the future, advanced technology such as electronic tagging may enhance further the traceability of livestock.

Bovine Spongiform Encephalopathy

In 1986, the UK agricultural industry was hit by a catastrophe – the discovery of Bovine Spongiform Encephalopathy, known simply as 'BSE' or 'mad cow disease'. Previous problems of market slumps and disease outbreaks were pushed into the shade by this discovery. The disease attacks the animal's nervous system, causing the animal to lose condition, become aggressive and lose control of its limbs, before becoming paralysed and finally dying. It was thought that the disease was linked to scrapie, a similar disease that affects the nervous system of sheep. It was also felt that it could have been passed on to cattle through infected meat and bone meal in feedstuffs. As a result, July 1988 saw MAFF introduce a ruminant feed ban. This had serious implications for rendering plants in Northern Ireland since much of their meat and bone meal was exported to GB. The importation of protein from ruminant species was banned in Northern Ireland in July 1988, though locally produced animal proteins in ruminant feed rations were not. Tough new regulations were also put in place to ensure that only animals from healthy herds were brought in. Despite the strictest precautions, BSE did reach Northern Ireland.

It was late November 1988 when BSE hit. The disease was diagnosed in a five-year-old homebred Friesian cow in County Down; the beast was immediately destroyed and the carcase incinerated. The Republic of Ireland discovered its first case only two months later, in January 1989. Almost 1,700 cases of BSE had already been confirmed in GB. In June 1988, MAFF had made BSE a notifiable disease and introduced a policy of compulsory slaughter, with compensation. Before the outbreak in 1988, the main priority for the farming community in Northern Ireland had been to keep their own herds free from the disease while maintaining exports. At this stage, there was no evidence to indicate that the disease could be transferred to the human population. Chief veterinary officer Bill Sullivan did, however, warn farmers that the disease had an incubation period of five years or more.

These beef carcases, which could have gone for human consumption, had to be destroyed due to BSE regulations.

Glenfarm Holdings

After the discovery of BSE in Northern Ireland in November, the Department implemented the same set of procedures as those in GB. Vernon Smyth, general secretary of the UFU, called on livestock producers not to panic. A second cow died in mid-December 1988 in County Fermanagh, and by now the numbers slaughtered in GB had risen to two thousand. In December, it was decided that a feed ban would take effect in January 1989. By March 1989, eleven Northern Ireland animals had fallen victim to BSE, and, by May, Northern Ireland's five education and library boards were discussing whether or not to ban beef from school dinners. Agriculture Minister John Gummer moved to calm fears about BSE, even having his daughter photographed, in May 1990, eating a hamburger.

Strong efforts were being made to assure the Albert Heijn supermarket group in the Netherlands, which was doing £15 million worth of business with Northern Ireland, that produce was as safe as anyone else's in Europe. Dr Ian Paisley MEP led the delegation that included representatives from the LMC, the Departments and Northern Ireland's main meat interests. They travelled to Amsterdam to spell out the message that there was nothing to fear from Northern Ireland meat. The problem of BSE was to have repercussions throughout the 1990s.

Livestock Improvement

The AI service run by the Department of Agriculture continued to have a major effect on improving the efficiency of both milk and beef production. In dairying, the need to look for improvement in quality and quantity of milk was stressed, together with the need for stock that 'wear' well. Use of the Canadian Holstein breed was having an increasing influence during the 1980s, as the percentage Holstein blood in the British Friesians rose steadily, although the Northern Ireland dairy herd was still 95 per cent Friesian. However, the introduction of milk quotas in 1984 resulted in a massive drop in dairy inseminations with an accelerated swing to the beef breeds.

In 1980, the AI service offered semen from Simmental, Charolais, Blonde d'Aquitaine and Limousin, in addition to the normal beef breeds – mainly Aberdeen Angus and Hereford. By 1985, Charolais semen had been available for twenty years and Simmentals for fourteen years. Limousins from 1978 and Blonde d'Aquitaines from 1985 were the most recent introductions to the AI service, with usage of the Limousin increasing substantially. In 1984–85, almost 20 per cent of beef inseminations were Limousin, although their growth rates were thought to be not as good as Charolais or Simmentals. However, in 1984–85, Simmental inseminations headed the popularity list, being responsible for almost 40 per cent of first inseminations by beef breeds. The popularity of the Charolais waned, probably because of its reputation for difficult calvings. Responsibility for providing a commercial AI service to cattle producers passed to private operators, AI Services NI, in 1988, under licence from the Department. Advertisements in March 1991 described the new operators as 'the complete cattle breeding service'. They claimed to offer an unequalled selection of dairy semen, top selection of beef bulls, an embryo transfer service and expert breeding advice.

The pros and cons for the use of AI in pigs were similar to those for use in cattle, and, in the late 1980s, initial trials were carried out on AI for sows. The main benefits included access to top-quality, performance tested boars and to different breeds, resulting in a genetically improved pig herd. One of the disadvantages in the early stages was that litter size and farrowing rates were slightly lower. Nevertheless, AI was being used widely and successfully in the pig herds of other European countries, such as the Netherlands and Denmark and, as the 1980s ended, it seemed likely that Northern Ireland would follow suit.

Grant Schemes and Environmental Issues

Two new grant schemes came into operation in October 1980, replacing and consolidating earlier schemes. They were the Agriculture and Horticulture Development and the Agriculture and Horticulture Grant Scheme. The former helped farming businesses to develop efficiently through an approved development plan, while the latter made available investment grants for works of a capital nature. The closing date for the schemes was mid-1985. The Agricultural Development Scheme started in January 1982, providing special aid for farmers in the LFAs of Northern Ireland. However, a new Agricultural Improvement Scheme came into operation in October 1985. This scheme required an improvement plan to be submitted for approval; the plan had to generate at least 1,100 hours work per year and demonstrate that earnings from the holding reached a certain level.

The Royal Society for the Protection of Birds called for an end to farm grants which, it claimed, were damaging moorland and ruining conservation measures. The society suggested that, in future, grants should be given to hill farmers only with the approval of the Department of the Environment's Conservation Branch. A report, *Hill Farming and Birds*, was made public in 1984; it proposed new management grants that would benefit both farming and conservation. The report also contained the society's first public endorsement of limited planning controls in the countryside. The society expressed alarm at the 'huge increase in peat extraction in Northern Ireland' which, it claimed, was destroying large areas of peatland, including moorland. Drastic measures were needed to protect moorland birds such as red grouse, marlins and lapwings whose haunts were being lost. By the mid-1980s, Northern Ireland had lost more than a quarter of moorland in forty years, with a consequent fall in bird populations.

Farming activities over many centuries had created the farming landscape and appearance of the Northern Ireland countryside, with its attractive mosaic of hedges and fields. However, some features such as natural woodlands, bogs and swamps, together with their particular wildlife, had almost disappeared. Only those pieces of land that were too difficult to cultivate were left untouched, and some of these were considered to be of great interest. During the 1980s, the Department of the Environment established a number of Areas of Special Scientific Interest (ASSIs) and more were being planned. By co-operating with farmers and landowners, the Department listed sites of scientific value that should be preserved. By the end of the 1980s, twelve ASSIs had been declared, covering a range of habitats. They included peat bogs, hazel woodlands, a large freshwater lake, a coastal site, limestone grassland and some old hay meadows that had never been fertilised. Farmers were, however, informed that restrictions would apply to certain notifiable operations which were considered to impact on the nature conservation interest. Examples included ploughing, drainage work, the application of slurry and the removal of stones.

The Department of Agriculture designated the first Environmentally Sensitive Area (ESA) in August 1988 – the Mourne and Slieve Croob ESA, which covered the foothills and lower slopes of the mountains. The landscape was very attractive, with its well-farmed fields enclosed by hedges and magnificent dry-stone walls. The objective was to encourage farmers to continue farming actively whilst, at the same time, improving the landscape and wildlife habitats. For the first time, farmers were offered financial incentives for the care and protection of the countryside. By the end of 1990, over 660 farmers in the ESA had entered into management agreements. An area of the Glens of Antrim was also designated as an ESA in July 1989 and, in the early 1990s, another followed in the west Fermanagh and Erne lakeland. The main items of work were repair and maintenance of dry-stone walls or hedges, erection of new walls and planting of new hedges and trees. Repairs of traditional buildings, gates and gateposts were also assisted under the scheme.

As far back as 1960, Sam Anderson of Magheralin wrote very descriptively:

And now it is April. There is a new carpet of green over field and meadow, and primroses and spring daisies peep from the margins and hedgerows. Lambs gambol and hares skip, and even the garrulous rook, busy with family affairs in a high grove has many companions for martin, swallow, swift, cuckoo and corncrake are arriving in quick succession.

How well he expressed his feelings for the countryside. Farmers, as custodians of the countryside, have an important part to play in caring for farmland and in maintaining rural areas as places of pleasure for urban visitors.

In the 1980s, however, the general public became more outspoken about the effects of farming methods on the environment. Any incidents of pollution by slurry and silage effluent that did occur were invariably given wide and critical coverage in the press. Intensive farming was continually under attack by vocal environmentalists. There was criticism of the cutting of hedges by machine and the erection of 'unsympathetic' farm buildings. Although the removal of hedges resulting in enlarged fields was regarded as good farming practice, many urban observers felt that this practice had gone too far. Their view was that trees played an important part in creating a pleasant environment, and they also provided shelter for animals, so why not retain them? In the late 1980s, farmers were starting to take their environmental role more seriously. Many undertook environmental improvement works such as planting hedges and trees, digging ponds and retaining areas of particular wildlife interest. The year 1989 was also designated 'Food and Farming Year'. Many events were organised throughout the UK and targeted at changing national attitudes to the agricultural industry.

Driving around the countryside, one became conscious of a new phenomenon which appeared in the early 1980s, often apparent in June or July – namely, damage to hawthorn hedges. Closer examination showed lengths of leafless hedge, covered with unsightly silken webbing. The small ermine moth was found to be doing the damage and it became more common throughout the 1980s and later. As the caterpillars grew, they fed and moved along the hedges, defoliating them as they went. Repeat attacks in the spring of each year damaged and weakened many thorn hedges throughout Northern Ireland, although by late summer the hedges were often back in leaf again.

Despite educational efforts and enforcement measures, the number of fatal accidents on farms remained high. There were eleven deaths during 1982 and thirteen in the following year. Tractors caused eight of the deaths over the two years. Electrocution and drowning were also among causes of death on the farm. In the fifteen years up until 1982, the average number of fatalities on farms was 14.6. Machines, farm animals and chemicals caused most farm accidents. Even in the 1990s, farm safety was still a problem. Walter Elliott, UFU president in June 1997, warned farmers that 'the silage season is a particularly dangerous time of the year and we must remind all workers to be aware of the dangers of working with machinery and equipment'. Despite many warnings over the years, there was clearly room for improvement in the general approach to farm safety.

Progress in Agricultural Education

In the 1980–81 academic year, 381 students, more than ever before, were attending courses at Northern Ireland's agricultural colleges. Many of the traditional courses at the three colleges were being offered on a part-time as well as a full-time basis. A new development in 1989 was the introduction of National Vocational Qualifications (NVQs) in agriculture. These were in line with education and training for industry in general. The new NVQs were made up of a number of stand-alone units covering different aspects of agriculture, such as health and safety, basic stock husbandry, feeding livestock and basic harvesting. Training focused on the essential skills or competences necessary to do the job. Students could choose a combination of units that best fitted their needs.

The diploma courses in agriculture did not stand still either. From September 1989, Greenmount offered a new course leading to the Higher National Diploma in agriculture. This three-year course, which included a year's work experience, was aimed at developing the skills needed in the modern agricultural industry. The course placed considerable emphasis on the management and marketing aspects of a farm business. It retained a strong practical approach, with husbandry and machinery skills being given an important role. Matt Boyd retired as Principal of Greenmount in 1984 after some thirty-six years of outstanding service to agricultural education. Many successful farmers in Northern Ireland are indebted to the thorough grounding in agriculture that they received at the college under his leadership.

One novel but progressive development took place in the certificate course at Loughry College in 1985. This was the introduction of a new non-farming subject called 'Personal and Social Development'. This was aimed at encouraging students to develop qualities, personal characteristics and abilities to cope with situations when dealing with people or organisations having contact with farmers. Communication and leadership skills were emphasised in the course, so that the students gained in confidence in preparation for their careers in a modern farming industry.

Changing Face of the Department

As the 1980s progressed, the face of the Department of Agriculture was changing too. The journal *Agriculture in Northern Ireland*, which had been produced for many years in the normal civil service A5 size and style, changed in size and format. It became an A4 document, with a revised presentation to attract readers and make its content more easily understood. This change complemented the restructuring of its advisory service. Because of the demand for more specialised advice, the Department reorganised its team of advisors into specialists in the main enterprises – dairying, beef and sheep, crops, pigs and poultry. Each county agricultural executive officer had overall responsibility for advisory work as well as administration of the various schemes in a particular county.

The Department had provided a 'free' advisory service to the farming industry for many years. In October 1988, however, it was announced that some services would no longer be free and that charges would be introduced. This brought Northern Ireland into line with the other countries in the British Isles. The Department offered a team of professional advisors, locally based, easily accessible and backed up by specialists and research personnel in the science service. It claimed that its advice was impartial, independent and confidential. Other advice was, of course, also available from the major commercial firms, although the degree of impartiality was a matter of opinion.

Ulster Farm in the 1980s

During 1980, under the leadership of Douglas Higginson, a radical change to the transport arrangements took place at Ulster Farm. A tipping platform was installed, enabling the unloading of a fleet of purpose-built raw-material semi-trailers. This dramatically reduced haulage charges – a saving that was further enhanced when outside hauliers took over the haulage of these trailers. The rendering equipment installed at Glenavy in the 1970s continued to give good service into the 1980s. By 1984, however, the rendering industry throughout Europe was beginning to adopt a different system. This made use of more energy-efficient technology through a system known as 'wet rendering'. This new system used large driers, fed from a pre-heating/pre-pressing/evaporation system. Liquid separation using tricanters took place after the pre-press stage, and waste heat from the process was reused at the evaporation stage. To this day, most of this equipment remains on site, incorporated in a much larger system that is fully compliant with the latest regulations. All of the equipment met and exceeded the standards stipulated in legislation by the Department of

Agriculture. At this stage, the number of company employees fell to about one third of that during the 1960s.

Ulster Farm and rendering generally were affected greatly by the identification of BSE in the late 1980s. The rendering process used in Northern Ireland was different from that used in GB, in that the hydrocarbon solvent extraction process was not used after 1973. During the 1980s, time/temperature combinations in Ulster Farm had been reduced but not to the same extent as had occurred in some of the plants in Britain. The transition period from batch to continuous processing using mainly Stord Bartz equipment began at Ulster Farm during 1982–84. The various types of offal were segregated and processed separately from 1988.

Syd Spence was appointed company secretary in June 1981, having been office manager since 1976. By 1982, he had restructured and computerised the accounting system and the shareholder register. This dispensed with the onerous task of handwriting cheques and dividend slips. Because of a rule change on age, Eddie Swain retired from the chairmanship in 1987 – it was truly the end of an era, as he had been at the helm since 1958. He was succeeded by Crawford Wadsworth, with Geoffrey Conn as vice-chairman.

Edward (Eddie) Swain, chairman of the Ulster Farm Board, 1958–87

James Swain

Conclusion

A glance back over the 1980s demonstrated how quickly circumstances could change. At the start of the 1980s there were no milk quotas, no ESAs and (as far as was known) no BSE. The Beef Variable Premium had gone, and the intervention system had been altered significantly. Farmers were encouraged to diversify away from commodities in surplus and to place greater emphasis on the environment. The public image of farming came under scrutiny in an unprecedented way, as animal welfare, food safety and environmental issues were queried endlessly by consumers and environmentalists. Consumers were also becoming more demanding than ever, and eating habits were changing, with snack meals and convenience foods on the increase. This was the background against which farmers were operating their businesses in the late 1980s.

The agricultural industry was still predominately grass-based, with over 70 per cent of total estimated receipts in the late 1980s coming from beef, sheep and dairying. However, as the 1980s closed, farmers were seriously concerned about how the BSE crisis would unfold and where it was all going to end. They wondered whether consumers would cease eating beef and whether beef production had a future at all. Pessimism abounded.

1990s
and beyond

Introduction

The 1990s opened with the cloud of BSE hanging over the agricultural industry – or the agri-food industry as it was increasingly being called. Little did anyone know that, before the end of 2001, another serious disease, FMD, would have added to the troubles of the industry. It eclipsed all other farming issues in that year. The BSE crisis was one event that had a very major effect on farming – some would say the largest impact since the Agriculture Act of 1947. The BSE problem reached a peak in March 1996 with the ban on feeding meat and bone meal and the decision to slaughter cattle over thirty months of age. These were major events that had a lasting impact on the industry. The MMB was wound up during the 1990s, and a farmers' co-operative created in its place. The PMB was also wound up. From the mid-1990s, Ulster Farm, then known as Glenfarm Holdings, expanded from rendering into other aspects of the food industry.

The increasing emphasis on the environment, which had been gathering momentum during the 1980s, continued towards the end of the century, adding to the industry's pressures. The large supermarkets that entered Northern Ireland were starting to wield some muscle with suppliers, and this was to have an impact at the end of the 1990s. Their demands increased for wholesome food that was free from residues. Food quality, animal health and consumers' reactions largely determined the production methods.

Up to 2002, the agriculture industry in Northern Ireland was still based primarily on family-run farms, with many farmers taking additional land in conacre to increase their acreage. Agriculture remained proportionally more important than in any other part of the UK. The Department has stated that only East Anglia contributes a greater proportion of regional Gross Domestic Product. Prior to the BSE crisis and the beef export ban (up to 1995), about two-thirds of agricultural output was sold outside Northern Ireland, mainly in GB. Farming incomes were low at the beginning of the 1990s but, despite BSE, they soon gathered pace, resulting in a significant improvement. The fundamental problem of food over-production remained within Europe. The European Union (EU) reformed the CAP and began the process of moving towards lower world prices. Quotas for milk remained in position during the 1990s.

Reforms within the European Union

The nine EC countries were joined by Greece in 1981 and by Portugal and Spain in 1986, making 'The Twelve'. In 1995 Austria, Sweden and Finland also became members. At the beginning of the 1990s there was much debate about the 'green pound' and its impact on the farming scene. Farmers complained that it was out of line with sterling and that this was affecting livestock returns. In February 1992, the member states signed the Maastricht Treaty. They agreed to work together in a grouping called the 'European Union' to review their political, economic and cultural links, including the idea of proceeding with a single currency.

Most of the EU budget was spent on agriculture, and there was pressure from other sectors to receive a fairer share. The aim of reforming the CAP was to bring the prices of EU agricultural commodities into line with world prices, and to reduce the persistent surpluses and their associated costs. The CAP reform proposals came from Mr Ray MacSharry MEP, Agriculture Commissioner during the 1990s and a hugely influential figure in agricultural politics. He was convinced that supply had to be brought more into line with demand. This was to be achieved by various means, such as setting aside farmland, continuing milk quotas and curbing beef production. The reform agreement was based on a reduction in those measures of support that influenced market prices, particularly the intervention arrangements. The overall impact of the CAP reforms in 1992, coupled with 'green pound' devaluations, was considered to be reasonably favourable for Northern Ireland.

The decision of the government to withdraw from the Exchange Rate Mechanism in September 1992 also had an immediate impact on Northern Ireland farming. The devaluation of sterling, which stemmed from this decision, gave a price boost to a wide range of farm products, although there were also increased import costs. The Single European Market was aimed at removal of the remaining barriers to trade within the EU, and the establishment of an internal market. On 1 January 1993, the twelve member states formally became one trading unit without internal frontiers. The much-talked-about Single Market had arrived. Frontiers within the EU disappeared, and with them all existing barriers to the free movement of animals and their products. From this date, Northern Ireland producers had virtually unrestricted access to European markets, presenting opportunities never before available. The adoption of common rules resulted in much freer movement of live animals and their products both into and out of Northern Ireland. It meant that Northern Ireland farmers had to compete increasingly with their counterparts in the Republic of Ireland, GB and mainland Europe. It was anticipated, however, that the wider market would also create opportunities.

Further reform to the CAP was agreed in 1999, described as 'Agenda 2000'. It was aimed at moving away from a market support policy towards direct income support, and encouraging rural development. At the time of writing, a mid-term review of 'Agenda 2000' is under way. The objective is to make the CAP more acceptable to consumers and taxpayers, particularly with the possibility of additional Eastern European member states within the EU.

An Integrated Administration and Control System (IACS), was adopted by the EU in an attempt to stamp out farm frauds. An IACS application became necessary in support of claims under Arable Crops, HLCAs, Beef Special Premium and Suckler Cow Premiums. Farmers were still concerned, however, about the amount of paperwork and bureaucracy which descended on them in the 1990s.

Further Implications of BSE

The development of legislation to stem the spread of BSE was still a priority as the 1990s began. The ban on the use of Specified Bovine Offal (SBO) in human food came into effect in Northern Ireland in January 1990. Compensation was being paid to farmers whose animals had to be slaughtered because of BSE. Although by mid-1990 Northern Ireland had had sixty-three cases of BSE, there was a feeling that the disease was under control. No homebred animal, for example, had contracted the disease since the introduction of the ruminant feed ban in 1989. With the ban on meat and bone meal and with BSE, it was no longer economical for rendering plants to collect fallen animals. Ken Maginnis, MP at Westminster, was concerned about the disposal of dead farm animals, which were no longer being collected by the by-products industry and were being dumped by the roadsides. In the spring of 1991, five months after rendering plants had announced their refusal to collect fallen animals, Agriculture Minister Jeremy Hanley announced that farmers would have an outlet for the disposal of fallen animals. As the service was a commercial one, costs would be involved, which, he believed, would work out at £80 per tonne or around £40 per animal. He reminded those who dumped dead livestock indiscriminately along the roadside that they would suffer severe fines if they transgressed.

Problems for Northern Ireland's beef industry increased in May 1990 when France banned all imports of beef and live cattle, resulting in the loss of valuable exports. Certain restrictions were imposed on imports of British beef into Germany and Austria. The loss of confidence in British beef products in Europe was mirrored by a similar loss of confidence within Britain itself. The future for the beef industry looked incredibly bleak. By 1993, fears were beginning to grow in the UK that the drive to eradicate BSE had failed. Several cases of the disease had also been found in homebred animals in Northern Ireland. One possibility was that contamination was occurring at rendering plants where there were no separate facilities for rendering SBO, and, in 1995, legislation was introduced in Northern Ireland which obliged rendering plants to have separate lines for dealing with SBO material.

In 1995, three people died of Creutzfeldt-Jakob Disease (CJD), a human transmissible spongiform encephalopathy, referred to in the press as the human form of 'mad cow disease'. A Belfast man, Maurice Callaghan, died from variant CJD in November. The disease was revealed in a new strain, and exposure to BSE was thought the most likely explanation. The LMC, however, assured consumers that there was no scientific or statistical evidence of a link between the cattle disease BSE and the human disease CJD. Unfortunately, such

Farmers protesting about BSE regulations at a rally at the King's Hall, Belfast, March 1996

Belfast Telegraph

reassurances could not be sustained and, on 20 March 1996, Health Minister Stephen Dorrell dropped a bombshell by announcing that it was likely that research would indicate a strong link between BSE and CJD. The BSE crisis had come to a head and there was little else to report about agriculture during 1996 because of BSE's complete dominance of the political agenda.

The immediate effects were a virtual halt in beef sales for a number of weeks, a dramatic drop in consumption across Europe, and the imposition of a ban on UK exports of beef and products containing beef or bovine material. The beef world throughout Britain reeled and has never been the same since. Hundreds of beef workers in Northern Ireland were laid off. All border roads were manned by gardaí to stop the transfer of cattle to the Republic. In schools, children were being offered a choice of beef-free meals; meat hauliers said that they were on the road to ruin; and the EU ruled that member states had acted legally in banning imports of British beef.

In a bid to restore consumer confidence the government decided in 1996 that a cattle slaughter scheme should be introduced, to ensure that all bovine animals over the age of thirty months at the time of slaughter should not enter the human food chain. This would involve a huge carcase-disposal programme that would take a considerable time to complete. Not surprisingly, many farmers went out of a business as a result of the BSE crisis – almost two thousand in the space of one year alone. Farmers felt that the government was not taking their plight seriously enough. Towards the end of 1996, South Down MP Eddie McGrady warned the House of Commons that the continuing BSE crisis could bring conditions to farming not unlike those precipitated by the Great Famine 150 years earlier. A further person, Andrew Hunter, died in Northern Ireland from CJD in April 2002. Since Health Secretary Dorrell's earth-shattering statement on that fateful March day, millions of pounds had been lost in export orders.

More Disease, including Foot and Mouth Disease

Tuberculosis remained a major problem in some areas of Northern Ireland in 1991, despite concerted control efforts. It was proving disappointingly slow to eradicate as strenuous efforts were made to ensure disease-free status. Northern Ireland had been officially brucellosis-free since 1982 but, during 1991, there had been eleven outbreaks of the disease in Counties Antrim, Armagh and Tyrone. Brucellosis therefore posed a continuing threat too. Sheep scab, a serious disease of sheep, had been controlled by having two compulsory dipping periods each year. This method of control was revoked in 1993 and the Department concentrated instead on controlling notified outbreaks of the disease.

With the creation of the Single European Market, the dangers of importing animal disease had increased. Frontier controls disappeared and the onus was on the member state of origin to ensure proper health status of animals being exported. Some states had been having trouble with the more serious diseases during the 1990s, so clearly there was a risk of introducing them into Northern Ireland. The transport of livestock with fast modern lorries meant that animals incubating disease might not show any signs until arriving at their destination.

But nothing prepared Northern Ireland farmers for the crisis that broke on 20 February 2001 with the discovery of FMD in England. The disease was confirmed in pigs sent for slaughter at an abattoir in Cheale, Essex. On the day after confirmation of the Cheale case, Agriculture Minister Bríd Rodgers MLA, agreed with MAFF – now the Department of Food and Rural Affairs – to ban the trade in live cattle, sheep, goats and pigs and their products between GB and Northern Ireland. Without doubt, this action protected Northern Ireland, illustrating graphically the importance of preventing disease entry. A 'Fortress

DARD staff disinfecting a lorry at Larne Port, as a precaution against foot and mouth disease

DARD

Farm' approach was the order of the day in an effort to keep disease at bay, as marts were shut down and livestock movements halted. There were only three outbreaks in Northern Ireland and one in the Republic of Ireland. There were, however, serious effects as intra-EU and third-country trade was banned from 21 February 2001.

It was a very difficult time for members of the Department's veterinary service and for everyone else connected with livestock farming. The service quickly undertook an exercise to trace all importations of cattle, sheep and pigs back to the beginning of the year. 'This was successful and correctly identified a load of sheep which had been imported for direct slaughter to a meat plant in Northern Ireland and had not been received by the plant,' Dr Bob McCracken, the chief veterinary officer, stated. Further investigations led to a farm at Meigh, south Armagh, where twenty-two imported sheep were examined. The animals proved negative for clinical signs of the disease, but, forty-eight hours later, they were showing different signs. Samples taken were submitted to the Surrey-based Foot and Mouth Reference Laboratory at Pirbright. The sheep were immediately slaughtered as a precaution. FMD was confirmed on 1 March 2001, and a three-kilometre protection zone and a ten-kilometre surveillance zone were established around the infected farm.

It is now acknowledged that the massive dissemination of FMD on the UK mainland occurred because the disease was not easily identifiable in sheep. Dr McCracken said that another factor was the absence of a twenty-day standstill on movements of flocks that had purchased infected sheep, allowing the disease to spread rapidly, especially through large sheep markets such as Longtown in Cumbria. The absence of sheep-movement licences also curtailed the efforts of government vets to trace the disease quickly and remove infected animals. Dr McCracken commented that:

> The uncontrolled and multiple movements of sheep in a two- to three-week period, when the disease had not been detected, was the single most important factor in contributing to the occurrence of over two thousand outbreaks during the epidemic. If sheep movement licences and a twenty-day standstill on movements had been imposed, the number of outbreaks in Great Britain might have been considerably less.

In the six weeks after 1 March 2001, no further cases were confirmed in Northern Ireland. A number of suspect cases were identified but samples proved negative. Because the disease was absent, the Department was able to make a case to the European Commission that Northern Ireland, with the exception of the Newry and Mourne District, should reopen for trade. From 3 April, then, for a brief period of time, trade in milk and milk products and in meat and meat products enjoyed a sense of normality. This was short-lived. Just before Easter, a 'hot' suspect was reported in the Coagh area and, on Good Friday, 13 April 2001, FMD was confirmed at another farm. The outbreak was quickly arrested but the next day the Department received a report of another suspect in Cushendall on a holding with cattle and sheep. The final outbreak was reported one week later on the edge of the one-kilometre cull zone around the Coagh outbreak.

In addition to the mandatory EU controls and a cull of all susceptible species within one kilometre of infected premises, licensing controls were also applied to the movement of all livestock in Northern Ireland. The only movements permitted outside the infected areas were of animals directly to slaughter. These strict controls were, in time, gradually relaxed to take account of welfare issues and to facilitate commercial trade. The Department undertook a massive serological survey of the sheep population to ensure that FMD was not circulating undetected. Around half a million sheep were sampled with no indication of a major problem. There is no doubt that the disease galvanised the farming community, the general public and the Department's staff into co-operating fully in every effort to contain the disease. However, on a UK-wide basis, the FMD outbreak again raised the question of whether the slaughter policy, rather than vaccination, had been the best means of dealing with the problem.

Livestock Production

The number of dairy cows increased from 274,000 in 1991 to 295,000 in 2001, despite the existence of milk quotas, although some quota had been leased from GB. Some farmers were prepared to pay unrealistic prices for quota and for land, but this may well have just been contributing to the already over-supplied milk market. Farmers were advised that a better approach would be to operate a high-output grassland system, with less land being required to carry the same stock. The released land could then be used for alternative enterprises such as cereals. In 2001, the census recorded 5,091 farms in Northern Ireland with dairy cows, resulting in an average herd size of almost fifty-eight cows, much larger than the comparable EU figure. Average milk yield per cow in 2001 was estimated as 6,220 litres per annum. In 2001, milk was contributing over 29 per cent of the agriculture industry's gross output.

Beef production was again the most common farming activity practised on Northern Ireland farms, as it had been for many years. According to the Department, over 80 per cent of farmers in the early 1990s had a beef production enterprise, although many herds were small. Farmers wondered, in 1990, whether to stay in beef production at all. They knew that, if they did, top-quality, healthy cattle would have to be produced without growth promoters, and in observance of the accepted animal welfare codes of practice. The animals would also have to be slaughtered in a humane way using hygienic processes – quite a demanding scenario! The production of beef from young bulls was developing in the early 1990s as a means of increasing efficiency. Experimental work at Hillsborough had shown that bulls would produce 20 per cent more saleable beef than steers slaughtered at the same age. The number of young bulls continued to rise, with 45,000 being slaughtered in 2001, an increase of 122 per cent over the previous year. The rise in the number of beef cows, which had begun in the late 1980s, continued, despite the shadow of BSE, reaching a peak of 345,000 in 1998 before falling back to 312,000 by 2001.

A major marketing initiative for beef was launched by the LMC in 1991 in the form of the Farm Quality Assurance Scheme. The objective was to

Free range and semi-intensive systems of egg production made a comeback in the 1990s.

Farm Trader

enhance the ability of the Northern Ireland beef industry to meet consumer requirements. The scheme was designed to provide customer assurance, to instigate a code of practice that retailers could use in promoting quality beef, and to allow the LMC to promote a product that was distinguishable by its scrupulous production methods. The code of practice covered all aspects of animal husbandry, welfare and environmental aspects of beef production at farm level. By 1993, Farm Quality Assured Beef from Northern Ireland was selling as a premium-quality product in supermarket stores throughout the Netherlands.

By June 2001, total pigs had fallen from 588,000 in 1991 to 386,000 in only 680 herds – the lowest figure for more than fifty years. As a means of providing customers with guarantees, the Ulster Curers' Association launched the Northern Ireland Pig Quality Assurance Scheme, in October 1993. Retailers were then able to purchase 'quality assured' pigmeat. By 1998, the pig industry had become depressed throughout the EU. This was exacerbated in Northern Ireland by a fire at a pig factory at Ballymoney, resulting in the loss of a significant proportion of processing capacity. This led to an even steeper fall in pig prices in 1998 and, by 2001, the value of finished pigs was £62 million – a considerable reduction from earlier years.

Sheep numbering 2.5 million were to be found on 10,500 farms in 2001, averaging about 120 ewes per flock. Up to 1993, there had been a long period of continual expansion in the breeding flock, followed by a levelling off and a slight decline by 2001.

Eggs and poultry meat were still the most concentrated enterprises on Northern Ireland farms at the end of the 1990s. Almost 99 per cent of the laying hens were found on only 106 farms in 2001, and all the broilers, almost nine million, were on 280 farms. The average laying flock was 1,700 hens, while the average broiler farm had over 31,000 birds in the June 2001 census. Although total poultry numbers were 19 per cent higher than ten years previously, the commercial laying flock had declined by 25 per cent. Public demand for non-cage and free-range eggs led, in the 1990s, to the development of several alternative systems of egg production. The systems included free-range and semi-intensive methods, both requiring open-air grass runs. The trouble was that the cost of producing eggs under any of these systems was much higher than with layers in cages. The imposition of welfare standards for laying hens may lead to the eventual banning of battery cages completely, even more spacious cages.

The average enterprise size on Northern Ireland farms in comparison with the UK, the Republic of Ireland and the EU is shown in Table 5.

Table 5: Average Enterprise Size in Late 1990s

Enterprise	Northern Ireland (2001)	United Kingdom (1999)	Republic of Ireland (1997)	EU 15 Countries (1997)
Dairy cows	58	72	33	25
Beef cows	19	27	12	15
Sheep	240	510	182	138
Pigs	568	585	694	114
Laying hens	1,727	1,403	128	189
Broilers	31,554	50,789	6,324	595
Cereals (hectares)	10.2	46.1	16.3	13.0
Potatoes (hectares)	6.3	10.6	2.1	1.4

Source: Statistical Review of Northern Ireland Agriculture 2001

Crop Production

As in previous decades, most farmland was under grass, with crops representing only 5 per cent of the total area. Almost 300,000 hectares of grass silage was made in 2001, the highest acreage ever. Grass, either grazed or conserved as winter feed, remained the basis for the ruminant livestock industry during the 1990s and beyond. The Department estimated in 1993 that grass contributed around 75 per cent of the total food requirement of dairy systems, 85 per cent of beef systems and 95 per cent of sheep systems. There was talk of techniques such as extending the grazing season in order to utilise more efficiently the potential of grassland.

With 33,000 hectares, barley, mostly spring sown, was the main crop grown, followed by potatoes at less than 7,000 hectares. Over the previous ten-year period, the acreage of cereals fell by 14 per cent, and potatoes by 38 per cent, continuing a downward trend in the overall cropped area, which had been apparent since the Second World War. Apples, mushrooms and vegetables were the main horticultural crops of significance at the end of the 1990s. The interest in organic foods that had begun in the 1980s showed a steady, albeit small, growth in demand during the 1990s. Some of the major food retailers showed limited interest, but the growth in sales in Northern Ireland was not on the same scale as in GB.

The Weather

It was a bleak winter for potato growers in 1992, when atrocious weather hit crops and brought losses estimated in the region of £6 million. John Hume MEP went to Brussels looking for extra aid from EU Commissioner Ray MacSharry, while his MEP colleague, Dr Ian Paisley, was out on the farms assessing farmers' problems at first-hand. The summer of 1995 was extremely 'long and hot', bringing drought throughout the UK. Reservoirs dried up and consumers were told to expect to pay the bill for one of the best summers on record. Potatoes were worst hit by the weather, with crops shrivelling in the ground. Hundreds of old wells and bore-holes were opened on farms as the authorities struggled to conserve dwindling reservoirs. The Meteorological Office warned that it would take two months of heavy rainfall to bring reservoir levels up to normal. Dairy farmers also struggled since their animals required a daily ration of around twenty gallons of water each. A Belfast produce merchant, Perry Donaldson, said that vegetable crops being grown for the winter had been badly hit. The price of potatoes was £330 a tonne, excellent for the producer, but anything but a bonus for the consumer. As rivers ran dry, fines for those convicted of pollution increased tenfold.

Farming Income and Debt

The 1990s did not have a good start in terms of farming income, with continuing worries about farmers' debts. There were also many external factors affecting the performance of farm businesses in 1990. These included the political changes going on in Eastern Europe, the disruption of the French lamb market, the Gulf crisis, high interest rates and rising inflation. Further uncertainty concerning the outcome of the General Agreement on Tariffs and Trade negotiations combined with the other factors to make 1990 a difficult year in which the income from almost every agricultural enterprise declined. In October 1990, it was estimated that farmers owed the banks a record £289 million on top of borrowings from other finance houses and lending institutions. Again, farm incomes were less than in GB, and prices for produce exported were lower because of Northern Ireland's peripheral location and transportation costs.

In 1997, the UFU Hill Farming Committee slammed government for failing to provide an extra £65 million for the following year's suckler cow scheme. According to committee chairman Harry Sinclair, average net farm income for hill farmers was less than £7,000 a year. There were warnings that farmers would take to the streets to make their feelings known and to draw attention to their difficulties. Members of NIAPA decided to step up action because of failed government promises. The chairman, Leslie Craig, said, 'The Government has failed us and we will have to do something.' The UFU sent representatives to take part in a 'Keep Britain Farming' protest in London, and NIAPA held a peaceful protest at Stormont. In the largest open-air gathering of farmers for many years, NIAPA laid on twenty-five buses and dozens of cars to bring one thousand farmers to the rally. Dr Ian Paisley, the Reverend William McCrea and many others addressed the meeting. Although it did allow farmers to let off steam, the meeting achieved little, and many farmers faced the prospect of bankruptcy.

A delegation of desperate farmers met Secretary of State Mo Mowlam to spell out the position of a once-prosperous industry. The delegation pointed to the huge drop in the number of dairy farmers, from 13,000 in 1973, to current numbers of just 6,000. The potato industry had also been halved in ten years. Bonfires burned across the UK, a symbol of farmers' protest at the government's agriculture policies. Northern Ireland's four main banks revealed that a 38 per cent drop in farm incomes had pushed debts up by £100 million since 1997. Estimates showed that debts to feed compounders had spiralled from £25 million in 1997 to £40 million in one year. Hire-purchase companies were owed £80 million in outstanding debts and were trying to recover the monies by restructuring a significant number of loans. The UFU president, Will Taylor, complained about the differences between farming in Northern Ireland and in the Republic: 'While farmers in the north were facing bankruptcy, those in the south were enjoying the benefits of the Celtic Tiger economy.' The Bank of England's decision to raise interest rates to 7.5 per cent before the summer was out was another blow.

Agricultural income fell in 1999 for the fourth year in succession. The Department concluded that the main reasons were the continuing impact of BSE and the beef export ban; the weakness of international agricultural markets; and the strength of sterling relative to the euro. The year 2001 saw the first significant rise in agricultural income since 1995. In comparison with GB, the outbreak of FMD in Northern Ireland required far fewer animals to be culled, and more valuable exports could be resumed at a time when products from GB were unavailable. For example, the higher-priced French lamb market was opened to Northern Ireland while British supplies were still restricted. Northern Ireland's 30,000 farm businesses had a gross output in 2001 of £1,190 million, compared with £998 million in 1991. During the 1990s, the Department produced regular reports on farm incomes in Northern Ireland, based on an annual farm-business survey. The surveys showed that the average cash income (the difference between cash receipts and expenditure) for a specialist dairy farm in 2000–01 was estimated at £28,400, while livestock farmers in LFAs were estimated to have received an average of £11,400. The average cash income for all farms in 2000–01 was estimated at £18,900 per farm.

The agriculture industry was, nevertheless, still supported to a significant level by public funds. The value of direct payments to farmers was £225 million in 2001, of which £164 million went to the cattle sector and £25 million to the LFAs for compensatory allowances. The comparable figure for 1991 for direct agricultural subsidies and grants was £99 million. This illustrates how significant the Exchequer input has been over the years; indeed, it has been significant over the past fifty years. Much of the credit for this must go to the Department of Agriculture and its predecessor departments, as well as to MAFF; all of these were involved, together with the farmers' representatives in the determination of government policy towards agriculture.

Revised Arrangements for Milk Marketing

Milk quotas had been operating for over seven years by the early 1990s. Yet in 1990–91, about one farmer in five suffered a superlevy penalty because of over-production, amounting in total to almost £1.5 million. Farmers were still debating whether to keep fewer cows, or more cows with a lower average milk yield. In seven years of milk quota, superlevy was imposed in all but three of those years. Fortunately, some inter-regional transfer of quota reduced the effect of the superlevy in Northern Ireland. Nevertheless, some farmers seemed to be gambling against the probability of a levy being imposed. The penalties were becoming more severe each year, reaching 25.25p per litre in 1991 with butterfat percentage a more critical factor than previously. Farmers were forced to give serious thought as to whether they could afford to take a penalty of over 25p per litre on excess milk.

Mr Robin Morrow, chairman of the MMB at the beginning of the 1990s, claimed that they had paid over £220 million to milk producers in 1991, despite a depressed market situation and a drop in milk supplies. He under-lined the importance of export markets as 70 per cent of Northern Ireland's production was sent out of the country. The average producer price in 1991 was 17.5p per litre, compared with 19p in 2001, although it had exceeded 24p per litre in 1995 and 1996.

During the 1990–91 period, the board became increasingly concerned that Strathroy Milk Marketing were threatening the stability of the liquid market through their milk purchasing activities. The board found it necessary to issue letters to Strathroy and those producers supplying it, again informing them of the illegality and consequences of their activities. Strathroy claimed in its advertisements, in July 1990, to have over two hundred producers, spread over Northern Ireland. The company was seeking more suppliers to meet its rising market demands. 'If you want more money for your milk, look no further than Strathroy,' said the advertisements in 1991. In December 1992, Strathroy claimed to be marketing milk for almost four hundred farmers, over 8 per cent of Northern Ireland production.

It was announced that deregulation of milk marketing in Northern Ireland would take place from 1 March 1995. The MMB became a voluntary co-oper-ative, relinquishing the statutory marketing arrangements it had enjoyed for many years. The new body, United Dairy Farmers, with a membership of over three thousand milk producers, was the largest dairy business in Northern Ireland. The co-operative collected the milk from farms and arranged for its delivery to a range of customers, mainly milk processors. Producers received a pooled milk price each month, which reflected the average return received from the processors, irrespective of how or where their milk supplies were used. The board's commercial arm, Dromona Quality Foods, which was pro-cessing about one-fifth of the milk produced in Northern Ireland, came into the new co-operative as a wholly owned but incorporated subsidiary. Milk supply to the co-operative in 1999–2000 was 922 million litres, and, together with its subsidiary companies, Dromona Quality Foods, Halib Foods International and United Feeds, it achieved a turnover of £216 million in that year.

Clenbuterol Abuse

During 1992, there were claims of misuse of the drug clenbuterol in beef pro-duction in some European countries, including both Northern Ireland and the Republic of Ireland. Clenbuterol, known as 'angel dust', was a veterinary drug for assisting with parturition – birth of calves – or treating respiratory problems in animals with pneumonia. However, when administered illegally to cattle at levels above the normal therapeutic dose, it had the effect of increasing muscle mass and decreasing body fat. Trials of the drug also revealed that it helped the conformation of lambs that had been injected with it. One could see how this could be an incentive to use clenbuterol illegally as a growth promoter. The problem was that there was considerable evidence that residues in cattle tissue posed a health hazard to consumers.

Some farmers appeared in court to answer charges of administering high doses of illegal drugs such as clenbuterol to farm animals. Clenbuterol was being given to cattle in animal feeds in the last four to six weeks before slaughter. The animals were said to show such a marked increase in meat gain that they were worth 10 to 20 per cent more when slaughtered. Department officials and veterinary officers closely monitored animals entering abattoirs, continually on the lookout for unusual characteristics, in an effort to eliminate clenbuterol abuse. Bill Hodges, Permanent Secretary of the Department of Agriculture, labelled users as 'fools and knaves', adding that they had 'no place in the industry', which, he said, they were 'wrecking'. Vigilant monitoring resulted in a steady decline in the number of cases of clenbuterol abuse detected in Northern Ireland.

Continuing Interest in Environmental and Welfare Issues

The environment continued to exercise the minds of politicians and the general public. Demands were coming from consumers for quality food products that had been produced in 'environmentally welfare-friendly conditions'. They wanted food produced from animals that had sufficient space and were not overstocked. The housing had to be designed so that it would neither cause injuries to the stock nor expose them to hazards such as adverse weather conditions, overcrowding, inadequate ventilation or noxious gases. Handling and husbandry of the animals were also important aspects to be considered. Welfare proposals covered the provision of insulation, heating and ventilation when planning new housing, to ensure proper air circulation, temperatures and relative humidity. With pigs, for example, there were many debates about loose housing systems, the provision of bedding, stocking rates and the most appropriate farrowing accommodation. Tethers and stalls were banned on pig farms from 1999.

Agriculture Minister Gillian Shepherd announced in 1994 that headage payments might be withheld where stocking density was so high that it could cause serious environmental damage. The trend with capital grant schemes changed too, with increasing emphasis being placed on preventing pollution and enhancing the countryside. An 'Environment Week' was held in May 1991, aimed at increasing the level of awareness about environmental matters. The Department of Agriculture co-operated fully, holding open days at Greenmount College and at Seskinore Forest and Game Farm. Topics covered included the effective use of fertilisers and slurry, containing and handling silage effluent, screening and painting of farm buildings, responsible use of

Donna Course, agronomist
with Ballymoney Foods,
examines some of the potato
crop of 2001 with Ian Hoy of
Cullybackey.

Brian Morrison Photography

pesticides, planting and maintaining hedgerows, organic production, environmental research and development, conservation and game management. This
represents a comprehensive list of the main environmental issues that farmers
had to take into account as they farmed in the 1990s. The theme of the environment also formed the main part of the Department's exhibit at Balmoral
Show in 1991, when these important issues were highlighted under the banner
of 'Food and the Countryside – The Way Ahead'. In July 1990, *Agriculture in
Northern Ireland* was produced for the first time from recycled paper.
(*Agriculture in Northern Ireland* continued until May 1996 when it was incorporated into *Farm Trader*, which was published by *Belfast Telegraph* Newspapers.)

The pilot ESAs established in the late 1980s were extended in the early
1990s, by which time they covered some 11 per cent of the agricultural land
area in Northern Ireland. Two further ESAs were designated in the Sperrins and
Slieve Gullion. People were becoming environmentally aware, with a growing
recognition that the countryside needed to be managed, and the best people
to do this were those who lived in it. It was they who could help to reduce
the dumps of old machinery and cars, polythene fertiliser bags and other
rubbish to be found spoiling the landscape. Conscious of policies at regional,
national and EU level, all urging environmental responsibility, farmers were
becoming countryside managers, whether they liked it or not.

During 1990, the Department carried out a wide-ranging review of the
economic and social needs of the most remote rural areas in Northern Ireland.
As a result, an independent Rural Development Council was established in
1991, representing a wide range of rural interests. This was funded by the
Department with assistance from the EC. The Rural Development Programme
was aimed at fostering the economic development and regeneration of rural
areas that would complement agriculture in the rural economy. About 690,000
people – over 40 per cent of Northern Ireland's population – live in rural

areas. The Rural Development Programme supported community-led regeneration projects that addressed the particular needs in some of Northern Ireland's most deprived rural areas.

The summer of 1990 saw the introduction of a new project called the 'Corncrake Hotline'. This was set up by the Royal Society for the Protection of Birds, which was asking people to telephone with reports of corncrakes. Farmers with breeding corncrakes on their land were eligible for a grant of £75 per hectare from the Department of the Environment if they were prepared to delay mowing until late July.

In the late 1990s, another 'food quality' issue made the headlines. This was referred to as genetically modified (GM) foods, where a single gene may be taken out or inserted by a process of genetic engineering. There were claims in 1999 that genetic modification could produce crops that were more resistant to pests, contained more vitamins, had better cooking qualities and had a longer shelf life. There were reports of yields of GM soya in the US being increased while using less herbicide. There were, however, also health-scare stories about GM foods causing flu outbreaks and cancer. It was claimed that rats fed GM potatoes had suffered damage to their immune systems and internal organs. The government confirmed in 2000 that there were a number of sites across GB where trials of GM oilseed rape crops were taking place. Scientists were saying that GM techniques could also be extended to farm animals, for example, to influence the protein content of milk. The debate about the human and health consequences of using GM techniques on farm crops and animals looks set to continue.

Marketing Northern Ireland Produce

There was continuing emphasis on promoting and increasing the market for produce from Northern Ireland. Bill Hodges, by now a prominent member of An Bord Bia, the Irish Food Board, argued in the late 1990s for increased efforts from all parties:

> We need to create a bigger market in Britain, Europe and even further afield for food from the island of Ireland. We sell on the same criteria, north and south. We both have the clean, green, environmentally friendly image and once a larger market is created it is up to individual firms to compete with each other as severely as they like for their share in that market. But let's get this bigger market first and worry afterwards who gets what share. We must strive to get more people to what Ireland, north and south, has to offer them.

Frozen chips produced by Ballymoney Foods on sale at Tesco

Phil Smyth

John Weatherup, commercial manager from Ballymoney Foods, tempts Agriculture Minister Bríd Rodgers to some freshly cooked chips on the company's stand at the 2002 IFEX event in Belfast.

Phil Smyth

The International Food, Drink and Catering Exhibition (IFEX) remained a major showcase for the Northern Ireland agri-food industry during the 1990s and into the 2000s. From the late 1980s, it had provided local companies with an excellent opportunity to promote their high-quality food with local and overseas buyers. It was interesting to note that an increasing number of food companies achieved the externally verified BS5750 or ISO9000 accreditation in the early 1990s. This confirmed the priority they were placing on total-quality assurance to remain competitive. Before he retired in 1994, Bill Hodges told the industry:

The challenge is there – the opportunity beckons. Will industry and government grasp the former and capitalise on the latter? I believe they must – there is no choice – and I am confident of a bright future for our greatest industry and its people.

Ulster Farm in the 1990s

Geoffrey Conn became chairman of Ulster Farm in 1991, with Robert Chesney as vice-chairman, a partnership that benefits the society to this day. Throughout the 1990s, under Geoffrey Conn's leadership, the society was guided through one of its most difficult decades – the pressures arising mainly from BSE.

Until the early 1990s, Ulster Farm was very much production oriented. A strategic review of the co-operative's future direction was undertaken in 1992. One of the first outcomes of the review was the purchase, in August 1993, of Ballymoney Foods. As the society still had considerable liquid assets, the decision was taken to widen its activities by acquiring synergistic enterprises. Michael Quinn was appointed, in 1994, as development executive. Among his first tasks was a review of the Ballymoney Foods operation and an

Philip Adams (centre), winner of the inaugural Ballymoney Foods Grower of the Year Award (2000), is pictured getting in the mood for a trip to Belgium with the Belgian national dish of mussels served with potato wedges.
Also pictured are Sam Adams (right) and Geoffrey Conn, chairman of Ballymoney Foods (left).

Pat McGuigan

investigation into the future strategy of the overall group. By 1995, it had been decided to restructure the organisation. This resulted in the renaming of the co-operative as Glenfarm Holdings and the transfer of the original business of rendering at Glenavy into a wholly owned subsidiary company, trading as Ulster Farm By-Products. It was recognised also that the existing small management team would require strengthening and, in 1995, Michael Quinn was appointed group chief executive of Glenfarm Holdings. This restructuring of the society was a key step, leaving it in a good position to deal with the BSE and FMD crises that eventually unfolded. Geoffrey Conn, Michael Quinn and Syd Spence all played key roles in developing a company which, until then, had enjoyed a relatively low profile as a very successful rendering plant. From 1995, professionally presented annual reports were circulated to all shareholders.

Douglas Higginson retired from the company in 1997, having worked for Ulster Farm for over thirty-eight years. Over that time, he had progressed from junior laboratory technician, through to production manager, works manager and finally to managing director. At every stage during that period, he had made a most significant contribution to the development and success of the company. Douglas had also represented Northern Ireland on the international stage, including a period as chairman of the UK Renderers Association.

The board was committed to developing the Glenfarm business by broadening its base and building it into a larger organisation, since the future in rendering in Northern Ireland alone offered no opportunity for growth. By 2001, the society had expanded considerably, then owning and controlling six

Pictured at the Northern Ireland Hospice in 2000, where Glenfarm Holdings was presented with a plaque to mark its donation of £380,000 to the Children's Hospice Appeal, are (from left) current chairman of the Children's Service, Brian Adgey, patron Gerry Kelly, chief executive of Glenfarm Michael Quinn, president of the Northern Ireland Hospice Paul Clarke, and chairman of Glenfarm, Geoffrey Conn.

Phil Smyth

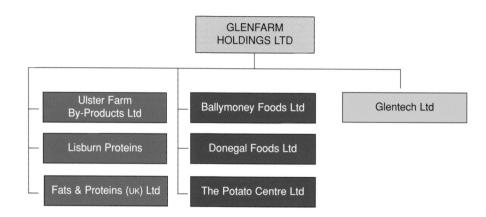

The structure of Glenfarm's
operating companies and
development company

Glenfarm Holdings

operating companies and one development company, Glentech. In addition to
Ulster Farm, the operating companies included Lisburn Proteins, Fats and
Proteins, Ballymoney Foods, Donegal Foods and The Potato Centre.

A new effluent plant, which was regarded as one of the best in Europe, was
installed at Glenavy in the early 1990s. This effluent facility continues to do a
very effective job today. The BSE crisis that struck UK agriculture in March
1996 instantly transformed all mammalian rendering businesses into waste-
processing operations geared to operate under government and EU animal
health and public health legislation. Ulster Farm co-operated fully with all relevant government agencies in facilitating the then necessary cull of cattle in
order to relieve the problems on farms as quickly as possible. Because of BSE,
the sale of mammalian meat and bone meal came to a halt in 1996 and tallow
could no longer be sold into its major market – animal feed. What had previously been feed material in effect became a waste product, the cost of which

Celebrating the formation of their
joint venture at Fats and Proteins
in Lancaster is Edmund Metcalfe,
managing director of Fats and
Proteins (left), shaking hands with
Harper Kilpatrick, general
manager of rendering at Glenfarm
Holdings. Also pictured are, from
left to right, Fats and Proteins site
manager Bob Harrison, process
manager John Parkinson, and
transport manager Mike Lowe.

Gerard Hearne

An aerial view of the
Ulster Farm plant
at Glenavy

Glenfarm Holdings

had eventually to be borne entirely by the industry. Compensation for loss of revenue was paid to the industry for a twelve-month period and phased out over the following twelve months, ending in March 1998. The direct loss of income was about £8.5 million, which, taken with a disposal cost of about £1 million, resulted in a cost to the livestock sector of about £9.5 million per annum.

In May 1997, Glenfarm purchased Lisburn Proteins, which, by then, was operating the only dedicated rendering plant in Northern Ireland dealing with all material arising from the compulsory slaughter and destruction of bovine animals over thirty months of age. This activity operated under contract and supervision of the Intervention Board (renamed the Rural Payments Agency from 2001).

In January 2000, Glenfarm entered into a joint venture with Mr Edmund Metcalfe to purchase Fats and Proteins, one of the leading rendering companies in GB. The potato-processing business of Ballymoney Foods expanded steadily during the 1990s. This development, which created new business, also saw the introduction of contract growing of potatoes for Ballymoney Foods, thus not only increasing the acreage of potatoes grown for the company, but also, for the first time, providing greater financial security for growers supplying the company. Donegal Foods, also involved in frozen-chip manufacturing, and located at Letterkenny in County Donegal, was acquired at the beginning of 2001.

Don McPherson (right) and
Freddie Carson, of Ulster Farm
By-Products, accept the
company's ISO14001
accreditation certificate from
Pauline Earl, of SGS Yarsley, the
international certification body.

Phil Smyth

During 1999, Ulster Farm achieved accreditation under the ISO
Environmental Management System – the first rendering company in Europe
to do so. Overall, the society progressed steadily throughout the 1990s,
recording a profit of £5.7 million on a turnover of £33.6 million in 2001.
This turnover was ten times that achieved by Ulster Farm nine years earlier.
At the fiftieth Annual General Meeting of Glenfarm Holdings in March 2002,
the chairman, Geoffrey Conn, reported that 'despite being one of the most
difficult years in farming history, Northern Ireland's leading agricultural
co-operative had continued to demonstrate successful growth through the
consolidation of its various business ventures'. The group's turnover for the
year ending 31 December 2001 increased by 51 per cent over the previous
year, while pre-tax profit increased by 7 per cent. Glenfarm's strategy of

Denis Sweeney of Donegal Foods
examines the year's harvest of
Rooster potatoes with a contracted
grower, Stewart McNutt of
Letterkenny.

John Caffrey

Senior management and current board members at Glenfarm. Front row, left to right: Claire Thompson (group head of finance), Syd Spence (group secretary), Geoffrey Conn and Robert Chesney. Back row, left to right: Noel Baxter, John Millar, Billy Martin, Crawford Wadsworth, John Liggett, Harper Kilpatrick (general manager – rendering) and John Weir.

Phil Smyth

expanding both its range of business activities and the geographic spread of the organisation was proving a success.

Referring to the impact of the serious FMD outbreak, Geoffrey Conn said:

From the first UK outbreak in February 2001, the impact of FMD on our rendering business has had major operational consequences. Fats and Proteins in Lancaster was totally dedicated to the processing of material arising from the cull from April to December 2001. And in Northern Ireland, Lisburn Proteins was nominated for the same purpose, but the service was only required for a three-week period last April.

He continued, 'The past twelve months has highlighted the necessity for efficiently run rendering facilities, and is testament to the fact that our business plays a vital role within today's agri-food industry.'

During 2001, Glenfarm transferred the Donegal Foods production of potato chips and consolidated all its potato-processing operations at the Ballymoney Foods site. Geoffrey Conn stated:

The concentration on a single site facility, together with a £4 million investment programme at Ballymoney, has enabled the company to dramatically increase production capacity and capability. As a result, Ballymoney Foods is now the sole frozen chip processing facility on the island of Ireland, and sales from both Ballymoney and Donegal Foods into retail and food service markets throughout GB and Ireland are increasing significantly. These achievements are also due to the strong links the group has developed with its potato suppliers, now sourced on an all-Ireland basis, ensuring the delivery of the highest quality products to the consumer.

Conclusion

By the end of the 1990s it was clear that with the current methods and systems, less land would be required in future to meet the requirements for food production. The long-term decline in the number of farms continued, with the average size increasing gradually. The trend towards fewer but larger farms has therefore continued relentlessly over the past fifty years. Advances in technology had improved productivity, and some products hovered on the verge of over-production. Action was taken, such as arable set-aside and the development of quotas on production. Farmers were giving more serious thought to alternative land-use options. Growing of trees was one longer-term option taken up by some farmers, but other options for alternative land use may emerge as the 2000s move on. The MMB, which had served the dairy farmers of Northern Ireland so well for so long, had gone, to be replaced by a co-operative, United Dairy Farmers. But the typical family-owned farm in Northern Ireland has displayed much strength. There was abundant grass on almost every farm; most farmers were very dedicated – they had a tradition of good stockmanship accompanied by high-quality stock; the countryside was considered to be relatively unpolluted, and consumers were keen to have the food products from Northern Ireland's farms. On the other hand, farmers had to pay dearly for some of their inputs, such as imported animal feed, and most farms were further from the main markets than their competitors. All of this emphasises how important the market has been and will remain for the products from Northern Ireland's farms.

Today, high-tech machinery and modern methods have replaced more traditional machinery and husbandry.

Michael Drake

The arrival in Northern Ireland, in the 1990s, of major multiple retailers, such as Tesco, J. Sainsbury and Safeway, was beginning to change relationships with farmers and food manufacturers. The large supermarkets had huge buying power, and their entry into the Northern Ireland market represented a very important challenge. There is little doubt that farmers and processors, if they are to survive, will have to be flexible and be prepared to produce food in line with the procedures and quality required by the large retailers. The food industry has, of course, been vital to the economy for many years. It remains the largest industry in Northern Ireland, employing directly almost 58,000 people in food production on farms in 2000. In the same year, a further 18,500 people were employed in the food-processing sector, while the manufacture and supply of inputs (such as animal feed, fertilisers and lime, medicines, farm machinery and services) employed 4,500 people. In that year, the processing sector achieved total external sales to GB, the Republic of Ireland and other foreign countries valued at £1,085 million.

Looking ahead, the successful farm business in the twenty-first century is likely to be one that has assessed what consumers really want, has prepared a sound business plan to produce it, manages the business using good husbandry techniques, and monitors progress on an almost daily basis. This approach of managing the farm as a business could, no doubt, have been applied to each one of the fifty years covered by this brief history.

A document published in 2001, entitled 'Vision for the future of the agri-food industry', was the outcome of deliberations by the Vision Group, representative of the agri-food industry in Northern Ireland, under the chairmanship of Peter Small of DARD. The group concluded that the vision should be:

A dynamic, integrated, innovative and profitable agri-food industry, focused on delighting customers in an evolving global market place and committed to developing its people. It will act as the guardian of our land-based heritage and rural environment and will help underpin and sustain the social fabric of rural areas.

It is a noble vision, which could prove a challenge to fulfil!

Conclusion

From 1950 onwards there has been a consistent trend towards fewer people working on the land and an increase in average farm size. Many farmers in Northern Ireland now carry out most of the work themselves, making use of other family or casual labour at peak times, and employing contractors for major activities such as combining of cereals or cutting hedges. This is a very different situation from that which pertained in the 1950s when farms were labour-intensive, involving many manual operations. Farm horses had begun to disappear from the fields. Mechanisation, including electrification of farms and increasing use of the tractor, contributed to the 'drift from the land'. This trend was, however, necessary to improve the efficiency and competitiveness of farming, and should not be regarded as a bad thing. Improved productivity per individual was essential to survival of the industry.

Two of the most significant advantages in Northern Ireland agriculture over the years have been the ability to grow high-quality grass, and the livestock husbandry skills of the farming community. The past fifty years have seen a big drop in the area of land being ploughed for crops, the 2001 figure being about one quarter of that for 1951. Potatoes and oats had reached a high level during the Second World War but reduced thereafter with decreased demand. Flax virtually disappeared from farms because of handling problems and declining prices. Barley acreage increased to take up some of the slack, but not to levels of the former tillage area. At the time, government was anxious to save on imports by encouraging the acreage of animal feed grown at home, in order to sustain intensive enterprises such as pig production. Cereal varieties more suited to Northern Ireland conditions and with better yields did help somewhat, but not sufficiently to alter the overall downward trend. Nevertheless, better weed and pest control contributed to increased yields over the years. There was also a huge swing from hay to silage from the 1950s onwards.

Over the years, animal production has become much more intensive, and farmers have tended to specialise in fewer enterprises. The quality of livestock has improved through artificial insemination and judicious importations of Continental breeds. Grassland has been managed more productively, and higher proportions of animals have been kept indoors. The total number of cattle has crept steadily upwards over the fifty years, but not to the same extent as sheep numbers, which have increased four-fold. Total numbers of poultry have fluctuated, but the numbers of laying birds have gone consistently downwards, while table chickens, referred to as broilers, virtually unknown in the 1950s, have increased steadily in number. Integrated groups involved in the provision of housing, equipment, birds and feed contributed to this trend. Similar integration principles were applied to other farm enterprises as time went on.

Government policy has had a major impact on Northern Ireland agriculture during the past fifty years. Through various price subsidies, production grants, capital grants and other employment and development grants, government has sought to assist and influence the direction of farming. The Northern Ireland Ministry of Agriculture (and its successor departments), the MAFF in Whitehall and the EEC all had a part in determining agricultural policy, and their contribution should be recognised. Subsidies and grants have often represented a significant part of farmers' incomes. The farm price review was for many years an important annual event during which commodity prices were hammered out. The UFU played an influential role here on behalf of its membership. Entry of the UK to the EEC in 1973 was to have a profound impact on farming, through acceptance of the CAP. It was aimed at improving the living standards of farmers while, at the same time, guaranteeing the supply of farm products at stable prices for consumers. However, problems arose when too much of some commodities were produced, causing surpluses. Milk quotas and set-aside schemes were introduced to bring supply more into line with demand. Various reforms to the policy have been, and are still being, made, and its impact will be felt in the years ahead.

Over the years, government made strenuous efforts to control animal disease in Northern Ireland. Yet diseases such as tuberculosis and brucellosis were a continuing problem from the 1950s, culminating with BSE in the 1990s and FMD in 2001. It would be difficult not to conclude that disease control has become more difficult as the Single European Market relaxed border controls, allowing a more liberal trading environment, and freer movement of livestock. In the early days, quarantine was the norm when importing livestock, but today, animals can reach Northern Ireland with relative ease, bringing with them the potential for new and serious disease. Towards the end of the fifty years, producing food in a way that did not harm the environment and which was acceptable from the animal-welfare point of view gained in importance. Consumers wanted quality food produced in housing conditions and stocking densities that took the welfare of the animals into account. Urban dwellers also wanted access to the countryside and began to express views on the environment which farmers cannot afford to ignore.

Throughout the past fifty years, attempts have been made by farmers to diversify, as pressure has come on certain enterprises. Mushroom production has been a success story, but other alternatives have not fared so well, and opportunities have been limited. Farming is now a business rather than a 'way of life'. Research findings have been applied successfully over the years, but farmers need to keep up-to-date. Agricultural education facilities in Northern Ireland have been second to none over the years, although uptake could have been improved. Better education of prospective farmers is an essential requirement if the industry is to survive.

Since 1950 agriculture has seen tremendous change, not only in how food

Quality food production at the Ballymoney Foods factory in County Antrim.

Glenfarm Holdings

commodities are produced but also in their marketing. Instead of the straight-forward production of milk, beef, pigs or poultry, farmers are now producing food as part of an overall 'agri-food' industry. The market is now much wider than in the 1950s and 1960s, when farming was aimed at satisfying the post-war demand for food within the UK, through a cheap food policy. The marketing boards of that period played a most useful role at the time. But the views of consumers have become more pronounced as time has gone on. Food technology is much more important than hitherto, and food processors will have to be innovative in the presentation of their products. Customers are now demanding more complete meals as opposed to individual items. They want more convenience foods and are anxious to ensure that their food is as wholesome and of as high a quality as possible. Those involved in the agri-food industry must accept that their future will be market-led and that concerns about food safety, animal welfare and the environment will have to be taken into account. The image of Northern Ireland has been of an area where food production is of a superior standard but, as in other parts of the UK, this image has been adversely affected recently because of BSE. Farmers will, therefore, have to give considerable priority to ensuring the highest level of consumer confidence in their products.

Farmers are now faced with a European and world market situation that is influencing the methods of production and the prices obtained. Co-operation within the production and marketing sectors through integrated groups has resulted in very efficient production of food. Over-production has taken place, resulting in the food surpluses of the 1980s and beyond. The problem at the beginning of the twenty-first century is that there is plenty of food, or even surpluses, worldwide, but the economics of distribution do not allow that food to be properly utilised. Many in the world are still starving. This problem will have to be addressed properly at international level in the future.

Ploughing in the Mourne
mountains in County Down.
The stone walls in the picture are a
distinctive aspect of the rural landscape
in parts of Northern Ireland.

Esler Crawford Photography

Appendix 1

Northern Ireland Livestock Numbers
June Census 1951 to 2001

Thousand head

	1951	1961	1971	1981	1991	2001
Dairy Cows	286	210	215	270	274	295
Beef Cows		98	254	205	254	312
Heifers in Calf	45	49	62	80	67	94
Other Cattle	630	717	853	881	938	979
Total Cattle	961	1,074	1,384	1,436	1,533	1,680
Breeding Ewes	335	586	477	575	1,204	1,232
Other Sheep	337	597	499	563	1,370	1,293
Total Sheep	672	1,183	975	1,139	2,574	2,525
Sows and Gilts	63	108	120	65	59	41
Other Pigs	522	925	1,037	562	529	345
Total Pigs	585	1,033	1,157	627	588	386
Laying Birds	16,974	9,163	8,121	4,348	2,748	2,143
Growing Pullets			3,177	1,366	946	735
Breeding Flock			585	651	964	2,145
Table Chickens		727	2,650	4,995	6,342	8,864
Turkeys, Geese and Ducks	863	324	131	125	292	461
Total Poultry	17,837	10,214	14,664	11,486	11,292	14,348
Horses and ponies	45	10	7★	★	6★	10★

★ Includes non-agricultural horses, not requested in previous censuses

Source: Ministry of Agriculture, Department of Agriculture for Northern Ireland and the Department of Agriculture and Rural Development: Statistical Reviews of Northern Ireland Agriculture

Appendix 2

Northern Ireland Crops and Grass
June Census 1951 to 2001

Thousand acres or hectares
Acres from 1951 to 1971; hectares thereafter (as in the respective census publications)

	1951 Acres	1961 Acres	1971 Acres	1981 Hectares	1991 Hectares	2001 Hectares
Oats	316.4	183.0	38.3	3.4	2.9	2.4
Wheat	1.3	4.9	2.5	0.4	5.9	4.1
Barley: Winter	2.9	111.0	140.0	3.1	5.0	2.8
Barley: Spring				47.9	32.0	30.0
Mixed Corn	4.5	2.0	10.4	1.0	0.2	0.2
Potatoes	144.1	75.0	42.1	12.5	10.8	6.7
Turnips, Swedes, Fodder Beet	12.5	4.2	1.0	0.6	0.6	N/A
Crop Silage	31.1	6.9	2.6	6.3	5.2	2.3
Other Field Crops						2.4
Total Crops	**512.8**	**387.0**	**236.9**	**75.3**	**62.6**	**50.9**
Fruit	10.1	9.0	7.7	2.4	1.9	1.5
Vegetables			2.4		1.3	1.5
Other Horticultural Crops					0.2	0.1
Total Horticultural Crops	**10.1**	**9.0**	**10.1**	**2.4**	**3.4**	**3.1**
Rotation Grasses	605.6	486.0	474.0	243.9	179.3	140.2
Permanent Grass	1,143.2	1,048.0	1,343.0	510.1	587.9	699.9
Total Grass	**1,748.8**	**1,534.0**	**1,817.0**	**754.0**	**767.2**	**840.1**
Total Crops and Grass	**2,271.7**	**1,930.0**	**2,064.0**	**831.7**	**833.2**	**894.1**

Source: Ministry of Agriculture, Department of Agriculture for Northern Ireland and the Department of Agriculture and Rural Development: Statistical Reviews of Northern Ireland Agriculture

Appendix 3

Persons Working on Northern Ireland Farms
1951 and 2001

1951		2001	
Owners Working on Farm:		Farmers and Partners:	
Male	52,073	Full Time	20,169
Female	6,063	Part Time	15,786
Total Owners	**58,136**	Total	35,955
Owners' Wives (if working on farm)	**24,043**	**Spouses of Farmers**	**6,520**
Other Family Workers		Other Workers	
Full-Time Family		Full Time	2,797
Male	24,640	Part Time	2,782
Female	11,393	Casual or Seasonal	8,308
Part-Time Family			
Male	5,583		
Female	3,772		
Total Family Workers	**45,388**		
Hired Workers			
Full-Time Hired			
Male	13,161		
Female	923		
Part-Time Hired			
Male	10,137		
Female	526		
Total Hired Workers	24,747		
Total Other Workers	70,135	Total Other Workers	13,887
Total Workers (including owners but excluding wives of owners)	128,271	Total Workers (excluding spouses of farmers)	49,842
Total Workers (including wives of owners)	152,314	Total Agricultural Labour Force	56,362

Source: Sixth Report upon the Agricultural Statistics of Northern Ireland 1930 to 1953; and Statistical Review of Northern Ireland Agriculture 2001

Appendix 4

Selected Agricultural Machinery and Equipment
1952 to 1984

	1952	1969	1984
Total Tractors	19,620	34,180	46,340
Tractor-drawn Ploughs	18,760	23,500	17,380
Rotavators	—	2,620	5,170
Combined Seed and Fertiliser and Corn Drills	1,710	2,190	1,840
Potato Planters	1,780	4,100	3,160
Mechanical Dung Spreaders	880	9,090	10,750
Slurry Tankers	—	1,530	6,700
Fertiliser Distributors	4,490	13,010	19,500
Tractor Mowers	11,870	22,000	20,520
Forage Harvesters	—	3,260	6,680
Greencrop Loaders	310		
Pick-up Balers	150	5,540	8,490
Binders	8,220		
Combine Harvesters	20	1,820	1,510
Potato Diggers	16,000	9,910	5,580
Grain Driers	—	890	670
Fore and Rear-end Loaders	200	6,320	10,800
Tractor Trailers (tipping)		8,530	21,700

Source: Seventh Report on the Agricultural Statistics of Northern Ireland 1952 to 1961; and Statistical Review of Northern Ireland Agriculture 1988

A dash indictates that the information was not available or not collected in that year.

A blank space indicates that the quantity was nil or insignificant in that year.

Appendix 5

Ulster Farm/Glenfarm Holdings Board History

The original Ulster Farm board was formed in 1952 and consisted of the following directors:

Name	Period	
J.M. Wadsworth	1952–1958	Chairman, 1952–1958
G. Ervine	1952–1988	Vice-chairman, 1958–1987
G.M. Fulton	1952–1974	
J.Martin	1952–1971	
A.E. Swain	1952–1988	Chairman, 1958–1987
H.W. West	1952–1959	
E. Wilson	1952–1968	
V. Wright	1952–1988	

The following directors served on the Ulster Farm board during the periods below:

Name	Period
Major W. Wilson	1959–1980
W.J. Chesney	1959–1979
W.G. McCollum	1975–1988
M. Ervine	1988–1989

The current Glenfarm Holdings board consists of the following directors:

Name	Period	
S.C. Wadsworth	1970–Present	Chairman, 1987–1991
C.G. Conn	1975–Present	Vice-chairman, 1987–1991
		Chairman, 1991–Present
J. Weir	1979–Present	
D.N. Baxter	1979–Present	
R.T. Chesney	1988–Present	Vice-chairman, 1991–Present
J.K. Millar	1988–Present	
J.McL Liggett	1988–Present	
W.R. Martin OBE	1989–Present	

Appendix 6

Key Personnel: Glenfarm Holdings Group

Name	Position	Period
William Gibson	Office Manager, Ulster Farm	1956–1957
	Company Secretary, Ulster Farm	1957–1976
Bill Hewitt	Factory Manager, Ulster Farm	1959–1962
	Managing Director, Ulster Farm	1962–1978
Douglas Higginson	Lab Technician, Ulster Farm	1959–1965
	Production Manager, Ulster Farm	1965–1974
	Works Manager, Ulster Farm	1974–1978
	Managing Director, Ulster Farm	1978–1997
Edmund Metcalfe	General Manager/Director, Fats and Proteins	1969–1995
	Managing Director, Fats and Proteins	1995–Present
Tony Addis	Lab Technician, Ulster Farm	1973–1988
	Production Manager, Ulster Farm	1988–1997
	Operations Manager, Ulster Farm	1997–2002
	Group Rendering Operations Manager, Glenfarm	2002
Ernie Hare	Engineer, Ulster Farm	1974–1993
William McLornan	Assistant Company Secretary, Ulster Farm	1974–1976
	Company Secretary, Ulster Farm	1976–1981
Syd Spence	Office Manager, Ulster Farm	1976–1981
	Company Secretary, Ulster Farm	1981–1995
	Group Secretary, Glenfarm	1995–Present
Ian Wallace	Managing Director, Ballymoney Foods	1978–1996
Ian Mairs	Works Manager, Ulster Farm	1979–1986
Don McPherson	Engineer, Ulster Farm	1994–1995
	Group Engineer, Glenfarm	1995–Present
Michael Quinn	Group Development Executive, Glenfarm	1994–1995
	Group Chief Executive, Glenfarm	1995–Present
Bob Harrison	Site Manager/Director, Fats and Proteins	1995–Present
Terry Webb	Group Accountant, Glenfarm	1995–1997
Harper Kilpatrick	Company Accountant, Ulster Farm	1996–1997
	Group Chief Accountant, Glenfarm	1997–2000
	General Manager – Rendering, Glenfarm	2000–Present
Mike Lowe	Logistics Manager, Fats and Proteins	1997–Present
Claire Thompson	Group Head of Finance, Glenfarm	2000–Present
David Harrison	Group Engineering Operations Manager, Glenfarm	2000–Present
Gary Millar	General Manager, Ballymoney Foods	2001–Present
John Weatherup	Group Marketing Executive, Glenfarm	2002

Select Bibliography

BARDON, JONATHAN, *Beyond the Studio: A History of* BBC *Northern Ireland* (Belfast, 2000)

Belfast News Letter, Farming Life (Belfast, 1963 to 2001)

Belfast Telegraph (Belfast, 1951 to 2001)

BINGHAM, S.P., *Capital Grants for Farmers: A Brief History* (London, 1983)

DEPARTMENT OF AGRICULTURE AND RURAL DEVELOPMENT, *Farm Incomes in Northern Ireland* (Belfast, 1996 to 2001)

— *Statistical Review of Northern Ireland Agriculture* (Belfast, 1999 to 2002)

— *Vision for the Future of the Agri-food Industry* (Belfast, 2001)

DEPARTMENT OF AGRICULTURE FOR NORTHERN IRELAND, *Statistical Review of Northern Ireland Agriculture* (Belfast, 1976 to 1998)

— *Ninth Report on the Agricultural Statistics of Northern Ireland, 1966–67 to 1973–74* (Belfast, 1977)

— *An Overview of the Northern Ireland Agri-food Industry* (Belfast, 1995 to 1998)

— *Northern Ireland Agricultural Statistics, 1984 to 1995* (Belfast, 1996)

DEPARTMENT OF FINANCE AND PERSONNEL, *Northern Ireland Annual Abstract of Statistics* (Belfast, 1982 to 2001)

EUROPEAN COMMISSION, *Europe Today* (London, 1997)

Farm Trader (Belfast, 1996 to 2000)

Farmweek (Portadown, 1951 to 2001)

FIRST TRUST BANK, *Business Outlook and Economic Review* (Belfast, 1992 to 1996)

— *Economic Outlook and Business Review* (Belfast, 1997 to 2001)

GENERAL REGISTER OFFICE AND NORTHERN IRELAND INFORMATION SERVICE, *Ulster Yearbook* (Belfast, 1951 to 1985)

GLENFARM HOLDINGS LIMITED, *Annual Report and Accounts* (Belfast, 1995 to 2001)

LIVESTOCK MARKETING COMMISSION, *Annual Report* (Belfast, 1967 to 1994)

MCCREARY, ALF, *On with the Show, 100 Years at Balmoral* (Belfast, 1996)

MAGEE, S.A.E., *Diversification on Northern Ireland Farms 1989* (Belfast, 1990)

MARKS, H.F., *A Hundred Years of British Food and Farming: A Statistical Survey* (London, 1989)

MILK MARKETING BOARD FOR NORTHERN IRELAND, *Annual Report* (Belfast, 1956 to 1994)

MINISTRY OF AGRICULTURE, FISHERIES AND FOOD, *Loaves and Fishes: An Illustrated History of the Ministry of Agriculture* (London, 1989)

— *The BSE Inquiry: Report, Evidence and Supporting Papers* (London, 2000)

MINISTRY OF AGRICULTURE FOR NORTHERN IRELAND, *Monthly Agricultural Report* (Belfast, 1951 to 1960)

— *Sixth Report upon the Agricultural Statistics of Northern Ireland, 1930 to 1953* (Belfast, 1957)

— *Seventh Report on the Agricultural Statistics of Northern Ireland, 1952 to 1961* (Belfast, 1967)

— *Eighth Report on the Agricultural Statistics of Northern Ireland, 1961–62 to 1966–67* (Belfast, 1970)

MINISTRY OF AGRICULTURE FOR NORTHERN IRELAND AND DEPARTMENT OF AGRICULTURE FOR NORTHERN IRELAND, *Agriculture in Northern Ireland* (Belfast, 1960 to 1996)

— *Outline of Northern Ireland Agriculture* (Belfast, 1963 to 1984)

— *At Your Service* (Belfast, 1972 to 1993)

NORTHERN IRELAND AGRICULTURAL TRUST, *Annual Report* (Belfast, 1968 to 1979)

TRUSTEE SAVINGS BANK NORTHERN IRELAND, *Business Outlook and Economic Review* (Belfast, 1985 to 1991)

ULSTER FARMERS' UNION, *The Farmers' Journal* (Belfast, 1951 to 1971)

— *The History of its First Seventy Years 1917 to 1987* (Belfast, 1989)

YOUNG, J.A., *The George Scott Robertson Memorial Lecture: Agricultural Progress and Future Trends in Northern Ireland* (Belfast, 1972)

Index

Page numbers in italics refer to illustrations.